National Vegetation Classification: Field guide to mires and heaths

National Vegetation Classification: Field guide to mires and heaths

T. Elkington, N. Dayton, D.L. Jackson and I.M. Strachan

This reprint edition published by Pelagic Publishing 2013
www.pelagicpublishing.com

ISBN-13 978-1-907807-50-3

This book is a reprint edition of ISBN-10 1-86107-526-X

Joint Nature Conservation Committee
Monkstone House
City Road
Peterborough
PE1 1JY
UK

This reprint edition published by Pelagic Publishing 2013
PO Box 725
Exeter
EX1 9QU

www.pelagicpublishing.com

ISBN-13   978-1-907807-50-3

This book is a reprint edition of ISBN-10   1-86107-526-X

Contents

1 Introduction

National Vegetation Classification

Since its development in the 1980s, the National Vegetation Classification (NVC) has become the standard classification used for describing vegetation in Britain. Whereas many other classifications are restricted to particular types of vegetation, the NVC aims to describe all the vegetation of Great Britain. This means that it is possible to analyse, and map, a complex site, composed of several habitat types (e.g. woodland, scrub, heathland and bog) using the same classification system.

The NVC is a 'phytosociological' classification, classifying vegetation solely on the basis of the plant species of which it is composed. The resulting communities can usually be correlated to other factors, such as geology and soils, climate, water chemistry and management; but the plant species alone are used to assign the vegetation to a community.

The NVC breaks down each broad vegetation type (e.g. heath, mire, woodland) into communities, designated by a number and name (e.g. H4 *Ulex gallii – Agrostis curtisii* heath, M10 *Carex dioica – Pinguicula vulgaris* mire, W4 *Betula pubescens – Molinia caerulea* woodland). Many (but not all) of these communities contain two or more sub-communities, designated by a letter (e.g. H4b *Ulex gallii – Agrostis curtisii* heath, *Festuca ovina* sub-community). Sub-communities in a few cases are further divided into variants (e.g. M10bi and ii).

Mires and heaths: scope of this guide

The second volume of *British Plant Communities* was published in 1991 (Rodwell 1991a). It provides a detailed account of 38 mire communities and 22 heath communities, giving information on their composition, structure and distribution, their affinities to other types of vegetation, both in Britain and on the Continent, and the relation of the communities described within the NVC to those previously described by other authors. The scope of this field guide is identical to that of Volume 2.

A number of vegetation types which might also be considered as 'mires' or 'heaths' are described in other volumes of *British Plant Communities*, and so are not included here. Aquatic, swamp and tall-herb fen communities can be found in Volume 4 (Rodwell 1995), whilst inundation communities,

dune slack communities and *Epilobium hirsutum* stands are described in Volume 5 (Rodwell 2000). 'Grass heaths', *Dryas* heaths and related lichen and bryophyte dominated vegetation are included in Volume 3 (Rodwell 1992), as are certain wet grasslands. Volume 1 (Rodwell 1991b) encompasses wet woodland and scrub vegetation. Companion guides to volumes 1 and 3 have also been published by JNCC (Hall *et al.* 2001; Cooper 1997).

Users of this guide should also note that most 'wet heath' vegetation is described in the NVC with the mires rather than the heaths (as M15 *Scirpus cespitosus – Erica tetralix* wet heath and M16 *Erica tetralix – Sphagnum compactum* wet heath), because of its floristic affinities. Helpful insight into the floristic relationships of NVC types can be gained from the Phytosociological Conspectus in Volume 5 of *British Plant Communities*. This places all NVC communities within a hierarchical framework of European vegetation.

Various gaps in coverage of the NVC have been identified at community and sub-community level subsequent to the publication of *British Plant Communities*. These include several mire and heath types, as outlined in JNCC Report No. 302 *Review of coverage of the National Vegetation Classification* (Rodwell *et al.* 2000). No attempt has been made to incorporate these here, pending further analysis and formal description.

Using this guide

The summary descriptions provided here are derived directly from the full accounts prepared by John Rodwell, but are in no way a substitute for them. Rather they are intended as an *aide-memoire* to assist surveyors in the field or for anyone else wishing to familiarise themselves with the overall scheme of classification for mires and heaths. Anyone who uses this book should always check their results against the frequency tables and full descriptions for each community in Volume 2 of *British Plant Communities*. The descriptions are not intended to take account of the results of recent survey work undertaken by the three country agencies (Countryside Council for Wales, English Nature, Scottish Natural Heritage) which may help circumscribe some of the communities more tightly and improve our understanding of community distributions.

A series of dendrograms have been produced to show the broad floristic relationships between the main communities and between the sub-communities for each community where these exist. These dendrograms are only intended as guides and should not be followed slavishly. Details of variants, if indicated, can be found in Volume 2 of *British Plant Communities*.

The amount of any particular species is referred to both in terms of its frequency and abundance. 'Frequency' refers to how often a plant is found in moving from one sample or vegetation to the next, irrespective of how much of that species is present in each sample. This is summarised in the published tables as classes denoted by the Roman numerals I to V: 1-20% frequency (that is, up to one sample in five) = I, 21-40% = II, 41-60% = III, 61-80% = IV, and 81-100% = V. The summary descriptions follow the usual convention of referring to species of frequency classes IV and V in a particular community as its constants, with those species of class III as common or frequent, of class II as occasional and of class I as scarce or rare. The term 'abundance', on the other hand, is used to describe how much of a plant is present in a sample, irrespective of how frequent or rare it is among the samples. It is summarised in the published tables as bracketed numbers for the Domin ranges, and is referred to in the text here, as in the published descriptions, using such terms as dominant, abundant, frequent and sparse.

The nomenclature for plant species used in *British Plant Communities* has been followed in this publication for consistency. Botanists more familiar with Stace's *New flora of the British Isles* (Stace 1997) may not recognise names such as (Stace equivalent in brackets): *Scirpus cespitosus* (*Trichophorum cespitosum*), *Carex demissa* (*C. viridula* ssp. *oedocarpa*), *C. lepidocarpa* (*C. viridula* ssp. *brachyrrhyncha*) and *Silene vulgaris maritima* (*S. uniflora*). Amongst cryptogams, the common lichen of heaths and bogs referred to here as *Cladonia impexa* is now generally known as *C. portentosa*.

References

Cooper, E A (1997) *Summary descriptions of National Vegetation Classification grassland and montane communities.* Joint Nature Conservation Committee, Peterborough (UK Nature Conservation No. 14).

Hall, J E, Kirby, K J and Whitbread, A M (2001) *National Vegetation Classification: field guide to woodland.* Joint Nature Conservation Committee, Peterborough.

Rodwell, J S, (ed) (1991a) *British Plant Communities. Volume 2. Mires and heaths.* Cambridge University Press, Cambridge.

Rodwell, J S, (ed) (1991b) *British Plant Communities. Volume 1. Woodlands and scrub.* Cambridge University Press, Cambridge.

Rodwell, J S, (ed) (1992) *British Plant Communities. Volume 3. Grasslands and montane communities.* Cambridge University Press, Cambridge.

Rodwell, J S, (ed) (1995) *British Plant Communities. Volume 4. Aquatic communities, swamps and tall-herb fens.* Cambridge University Press, Cambridge.

Rodwell, J S, (ed) (2000) *British Plant Communities. Volume 5. Maritime communities and vegetation of open habitats.* Cambridge University Press, Cambridge.

Rodwell, J S, Dring, J C, Averis, A B G, Proctor, M C F, Malloch, A J C, Schaminée, J N J and Dargie, T C D (2000) Review of coverage of the National Vegetation Classification. *JNCC Report No. 302.*

Stace, C (1997) *New flora of the British Isles.* 2nd ed. Cambridge University Press, Cambridge.

2 Dendrogram keys to mire communities

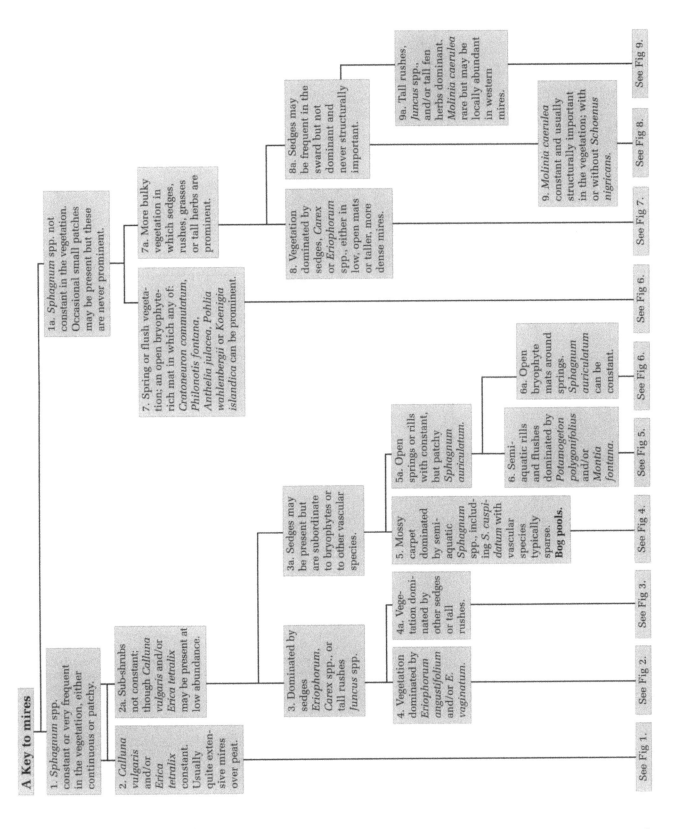

A Key to mires

1. *Sphagnum* spp. constant or very frequent in the vegetation, either continuous or patchy.

1a. *Sphagnum* spp. not constant in the vegetation. Occasional small patches may be present but these are never prominent.

2. *Calluna vulgaris* and/or *Erica tetralix* constant. Usually quite extensive mires over peat.

2a. Sub-shrubs not constant; though *Calluna vulgaris* and/or *Erica tetralix* may be present at low abundance.

3. Dominated by sedges *Eriophorum*, *Carex* spp., or tall rushes *Juncus* spp.

3a. Sedges may be present but are subordinate to bryophytes or to other vascular species.

4. Vegetation dominated by *Eriophorum angustifolium* and/or *E. vaginatum*.

4a. Vegetation dominated by other sedges or tall rushes.

5. Mossy carpet dominated by semi-aquatic *Sphagnum* spp., including *S. cuspidatum* with vascular species typically sparse. **Bog pools.**

5a. Open springs or rills with constant, but patchy *Sphagnum auriculatum*.

6. Semi-aquatic rills and flushes dominated by *Potamogeton polygonifolius* and/or *Montia fontana*.

6a. Open bryophyte mats around springs. *Sphagnum auriculatum* can be constant

7. Spring or flush vegetation; an open bryophyte-rich mat in which any of: *Cratoneuron commutatum*, *Philonotis fontana*, *Anthelia julacea*, *Pohlia wahlenbergii* or *Koenigia islandica* can be prominent.

7a. More bulky vegetation in which sedges, rushes, grasses or tall herbs are prominent.

8. Vegetation dominated by sedges, *Carex* or *Eriophorum* spp., either in low, open mats or taller, more dense mires.

8a. Sedges may be frequent in the sward but not dominant and never structurally important.

9. *Molinia caerulea* constant and usually structurally important in the vegetation; with or without *Schoenus nigricans*.

9a. Tall rushes, *Juncus* spp., and/or tall fen herbs dominant. *Molinia caerulea* rare but may be locally abundant in western mires.

See Fig 1.

See Fig 2.

See Fig 3.

See Fig 4.

See Fig 5.

See Fig 6.

See Fig 6.

See Fig 7.

See Fig 8.

See Fig 9.

9

Mires Figure 1

Fig 1. Key to vegetation with constant *Sphagnum* spp. and constant ericoid sub-shrubs

1. *Eriophorum vaginatum* (either as prominent tussocks or sparse fronds) and/or *Sphagnum papillosum* and/or *S. magellanicum* constant. Deep peat > 1 m. **Blanket and raised mires**.

1a. See next page

2. *Molinia caerulea* and/or *Scirpus cespitosus* constant with at least some: *Narthecium ossifragum*, *Eriophorum angustifolium*, *Potentilla erecta*, *Sphagnum papillosum* and *S. capillifolium*.

2a. *Scirpus cespitosus* not constant and rarely prominent, though can be frequent in patches. *Molinia caerulea* generally absent (can form mosaics with tussocky *Eriophorum vaginatum* in degraded forms of M19.)

3. *Sphagnum* spp. prominent, especially *S. papillosum*, *S. capillifolium* and *S. tenellum*, often *S. magellanicum*. Associates include at least some of *Vaccinium oxycoccos*, *Drosera rotundifolia* and *Odontoschisma sphagni*.

3a. *Sphagnum* spp. patchy, though *S. capillifolium* constant. *S. tenellum*, *S. magellanicum* and *Odontoschisma sphagni* generally absent or very rare.

4. *Calluna vulgaris* and *Eriophorum vaginatum* form the bulk of the vegetation, *E. vaginatum* usually tussocky, with *Sphagnum capillifolium*, *E. angustifolium* and pleurocarpous mosses: *Pleurozium schreberi*, *Hylocomium splendens* and *Hypnum jutlandicum*.

4a. Vegetation dominated by tussocks of *Eriophorum vaginatum* with *Calluna vulgaris* and *Sphagnum* spp. at best infrequent and patchy. Mounds of *Empetrum nigrum* and/or *Vaccinium myrtillus* with constant *Eriophorum angustifolium* and *Deschampsia flexuosa*.

M17 *Scirpus cespitosus – Eriophorum vaginatum* blanket mire

M18 *Erica tetralix – Sphagnum papillosum* raised and blanket mire

M19 *Calluna vulgaris – Eriophorum vaginatum* blanket mire

M20b *Eriophorum vaginatum* blanket and raised mire, *Calluna vulgaris – Cladonia* spp. sub-community

Burning can increase the proportion of *Calluna vulgaris* or *Deschampsia flexuosa* and *Vaccinium myrtillus*. Grazing usually decreases the abundance of *C. vulgaris*, ultimately to its disappearance, especially if combined with large-scale burns. Such disturbance can shift the floristics of this community towards the impoverished M20 mire. Lower altitude stands (<400 m) to the west usually include constant *Erica tetralix* which is replaced in higher and more eastern stands by *Empetrum nigrum* and *Rubus chamaemorus*.

The *Eriophorum* mire is apparently biotically derived from M19 blanket bog or M18 raised mire; the degree of floristic impoverishment depending on the intensity and duration of destructive management practice. There is therefore a continuous gradation between richer and poorer stands. In the wet north-west some *Sphagnum* spp. can persist even in highly degraded stands, especially *S. recurvum* which is relatively tolerant to some disturbance.

From previous page:
1a. Both *Eriophorum vaginatum* and *Sphagnum papillosum* generally absent, though may be patchily present at low frequency in the wetter sub-communities of the M15 *Erica – Scirpus* mire.

5. *Calluna vulgaris* and *Erica tetralix* both prominent with *Eriophorum angustifolium* and *Molinia caerulea* constant.

5a. *Erica tetralix* constant. *Calluna vulgaris* absent or present at low frequency in the vegetation.

6. *Narthecium ossifragum* constant and abundant with *Drosera rotundifolia*, *Eriophorum angustifolium*, *Sphagnum papillosum*, plus *S. auriculatum* and/or *S. recurvum*. *Schoenus nigricans* absent or very rare.

6a. *Narthecium ossifragum* present at low frequency. Ericoid sub-shrubs and at least one of *Scirpus cespitosus*, *Molinia caerulea* and *Eriophorum angustifolium* make up the bulk of the vegetation over patchy sphagna, *S. capillifolium* or *S. compactum*. **Wet heaths.**

7. *Erica tetralix* usually the most prominent sub-shrub. *Sphagnum* layer dominated by *S. compactum* and *S. tenellum*. *Molinia caerulea* usually dominant amongst the vascular monocot associates.

7a. *Calluna vulgaris* usually the most prominent sub-shrub except where grazing pressure has favoured the expansion of *E. tetralix*. *Sphagnum* spp. dominated by *S. capillifolium* and *S. subnitens* with occasional *S. papillosum* in the wetter sub-communities.

8. *Schoenus nigricans* constant with prominent *Narthecium ossifragum* and *Sphagnum subnitens*.

8a. *Molinia caerulea* dominant, dense and tussocky. Sparse *Erica tetralix* and *Potentilla erecta* constant. *Sphagnum palustre* and *S. recurvum* can be prominent in wet north-western stands.

M21 *Narthecium ossifragum – Sphagnum papillosum* valley mire

May be confused with the flushed wet heath, M15a; however the sedges *Carex panicea*, *C. echinata* are constant in the latter and *Myrica gale* is more frequent. *Pinguicula vulgaris* is common in M15a but is replaced in the M21 community by less frequent *P. lusitanica*.

M16 *Erica tetralix – Sphagnum compactum* wet heath

Wet heath primarily of the south and east of Britain, this community provides the major locus for *Scirpus cespitosus* and *Eriophorum vaginatum* (though rare) in this part of the country.

M15 *Scirpus cespitosus – Erica tetralix* wet heath

This is a very variable vegetation type and, of the major components, any can be dominant with up to two of the others missing. A community primarily of north-west Britain, it includes most of the heather-dominated vegetation intermediate in character between the dry heath and blanket mire types. Where heavily grazed and/or burnt, the ericoids can become very sparse and these stands are often transitional to the *Molinia*-dominated community M25, or drier stands to the *Juncus squarrosus* grassland U6.

M14 *Schoenus nigricans – Narthecium ossifragum* mire

M25a *Molinia caerulea – Potentilla erecta* mire, *Erica tetralix* sub-community

Burning and grazing of M15 wet heath can favour an increase in *Molinia* which shifts the vegetation towards that of the *Erica tetralix* sub-community of M25. The similarities between M25a and *Molinia*-rich M15 show the transition between these two communities through disturbance.

Mires Figure 2

Fig 2. Key to vegetation with constant *Sphagnum* spp. and dominated by *Eriophorum vaginatum* and/or *E. angustifolium*

1. Vegetation dominated by *Eriophorum angustifolium* in which *Sphagnum cuspidatum* and/or *Drepanocladus fluitans* can be prominent. Bog pools and erosion hagg runnels within blanket peat.

1a. Vegetation dominated by *Eriophorum vaginatum* in which *E. angustifolium* can be constant but is always subordinate to *E. vaginatum*.

2. Tussocky *E. vaginatum* with very patchy *Sphagnum capillifolium*, constant *Deschampsia flexuosa* and mounds of *Vaccinium myrtillus* and/or *Empetrum nigrum nigrum*. *Calluna vulgaris* usually absent but can be present as scattered sprigs at low frequency. North-western stands also feature prominent *Polytrichum commune*.

2a. It can be difficult to separate richer forms of the M20b sub-community from poor and degraded stands of the M19 blanket mire, since there is a continuum of vegetation types between these two communities with increasing disturbance to the latter. Generally, if *C. vulgaris* is constant with a reasonably intact *Sphagnum* flora this community can be categorised with the M19 mires, though *E. vaginatum* may be visually dominant.

M3 *Eriophorum angustifolium* bog pool

M20b *Eriophorum vaginatum* blanket and raised mire, *Calluna vulgaris – Cladonia* spp. sub-community

M19 *Calluna vulgaris – Eriophorum vaginatum* blanket mire

This community is typical of recent or disturbed bog pools or forms an early seral stage in the transition from exposed peat back to mire vegetation. Poorer stands of the M2b *Sphagnum recurvum* bog pool can be very similar to *Sphagnum*-rich M3 stands, reflecting the transitional continuum between these communities.

Mires Figure 3

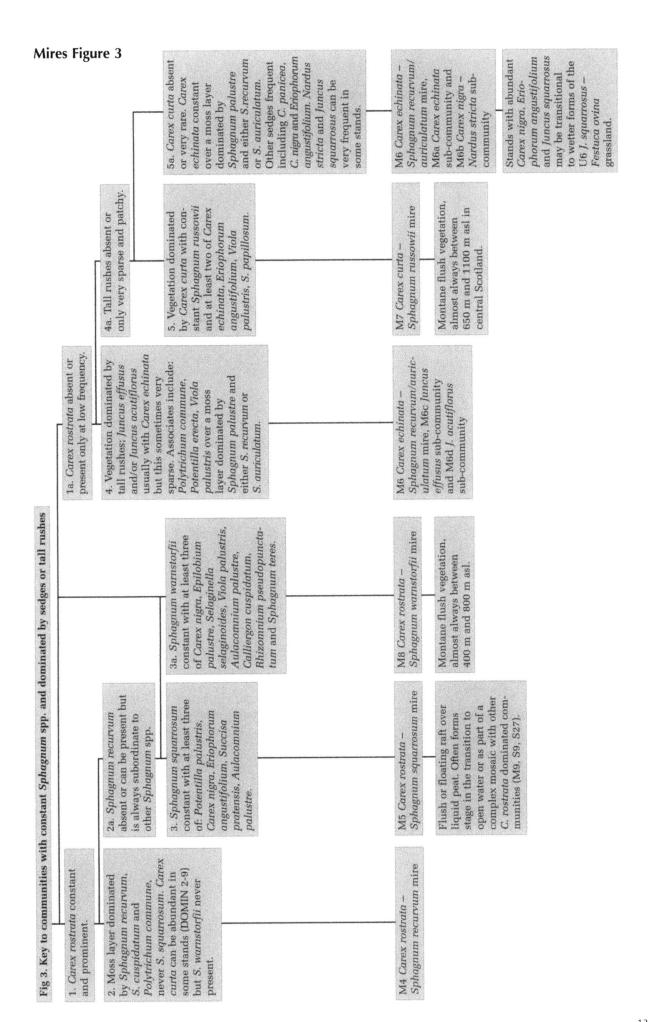

Fig 3. Key to communities with constant *Sphagnum* spp. and dominated by sedges or tall rushes

1. *Carex rostrata* constant and prominent.

1a. *Carex rostrata* absent or present only at low frequency.

2. Moss layer dominated by *Sphagnum recurvum*. *S. cuspidatum* and *Polytrichum commune*, never *S. squarrosum. Carex curta* can be abundant in some stands (DOMIN 2-9) but *S. warnstorfii* never present.

2a. *Sphagnum recurvum* absent or can be present but is always subordinate to other *Sphagnum* spp.

3. *Sphagnum squarrosum* constant with at least three of: *Potentilla palustris, Carex nigra, Eriophorum angustifolium, Succisa pratensis, Aulacomnium palustre.*

3a. *Sphagnum warnstorfii* constant with at least three of *Carex nigra, Epilobium palustre, Selaginella selaginoides, Viola palustris, Aulacomnium palustre, Calliergon cuspidatum, Rhizomnium pseudopunctatum* and *Sphagnum teres.*

4. Vegetation dominated by tall rushes; *Juncus effusus* and/or *Juncus acutiflorus* usually with *Carex echinata* but this sometimes very sparse. Associates include: *Polytrichum commune, Potentilla erecta, Viola palustris* over a moss layer dominated by *Sphagnum palustre* and either *S. recurvum* or *S. auriculatum.*

4a. Tall rushes absent or only very sparse and patchy.

5. Vegetation dominated by *Carex curta* with constant *Sphagnum russowii* and at least two of *Carex echinata, Eriophorum angustifolium. Viola palustris, S. papillosum.*

5a. *Carex curta* absent or very rare. *Carex echinata* constant over a moss layer dominated by *Sphagnum palustre* and either *S.recurvum* or *S. auriculatum.* Other sedges frequent including *C. panicea, C. nigra* and *Eriophorum angustifolium. Nardus stricta* and *Juncus squarrosus* can be very frequent in some stands

M4 *Carex rostrata* – *Sphagnum recurvum* mire

M5 *Carex rostrata* – *Sphagnum squarrosum* mire

Flush or floating raft over liquid peat. Often forms stage in the transition to open water or as part of a complex mosaic with other *C. rostrata* dominated communities (M9, S9, S27).

M8 *Carex rostrata* – *Sphagnum warnstorfii* mire

Montane flush vegetation, almost always between 400 m and 800 m asl.

M6 *Carex echinata* – *Sphagnum recurvum/auriculatum* mire, M6c *Juncus effusus* sub-community and M6d *J. acutiflorus* sub-community

M7 *Carex curta* – *Sphagnum russowii* mire

Montane flush vegetation, almost always between 650 m and 1100 m asl in central Scotland.

M6 *Carex echinata* – *Sphagnum recurvum*/ *auriculatum* mire, M6a *Carex echinata* sub-community and M6b *Carex nigra* – *Nardus stricta* sub-community

Stands with abundant *Carex nigra, Eriophorum angustifolium* and *Juncus squarrosus* may be transitional to wetter forms of the U6 *J. squarrosus* – *Festuca ovina* grassland.

13

Mires Figure 4

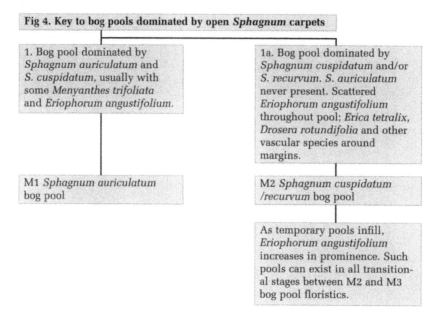

Fig 4. Key to bog pools dominated by open *Sphagnum* carpets

1. Bog pool dominated by *Sphagnum auriculatum* and *S. cuspidatum*, usually with some *Menyanthes trifoliata* and *Eriophorum angustifolium*.

1a. Bog pool dominated by *Sphagnum cuspidatum* and/or *S. recurvum*. *S. auriculatum* never present. Scattered *Eriophorum angustifolium* throughout pool; *Erica tetralix*, *Drosera rotundifolia* and other vascular species around margins.

M1 *Sphagnum auriculatum* bog pool

M2 *Sphagnum cuspidatum /recurvum* bog pool

As temporary pools infill, *Eriophorum angustifolium* increases in prominence. Such pools can exist in all transitional stages between M2 and M3 bog pool floristics.

Mires Figure 5

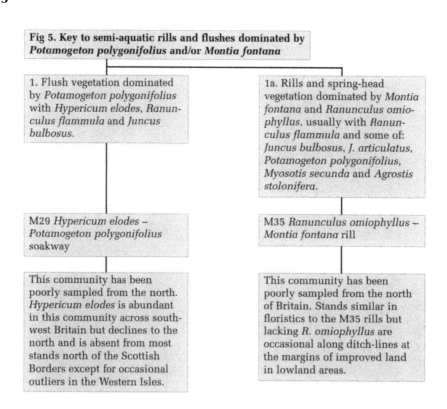

Fig 5. Key to semi-aquatic rills and flushes dominated by *Potamogeton polygonifolius* and/or *Montia fontana*

1. Flush vegetation dominated by *Potamogeton polygonifolius* with *Hypericum elodes*, *Ranunculus flammula* and *Juncus bulbosus*.

1a. Rills and spring-head vegetation dominated by *Montia fontana* and *Ranunculus omiophyllus*, usually with *Ranunculus flammula* and some of: *Juncus bulbosus*, *J. articulatus*, *Potamogeton polygonifolius*, *Myosotis secunda* and *Agrostis stolonifera*.

M29 *Hypericum elodes* – *Potamogeton polygonifolius* soakway

M35 *Ranunculus omiophyllus* – *Montia fontana* rill

This community has been poorly sampled from the north. *Hypericum elodes* is abundant in this community across south-west Britain but declines to the north and is absent from most stands north of the Scottish Borders except for occasional outliers in the Western Isles.

This community has been poorly sampled from the north of Britain. Stands similar in floristics to the M35 rills but lacking *R. omiophyllus* are occasional along ditch-lines at the margins of improved land in lowland areas.

Mires Figure 6

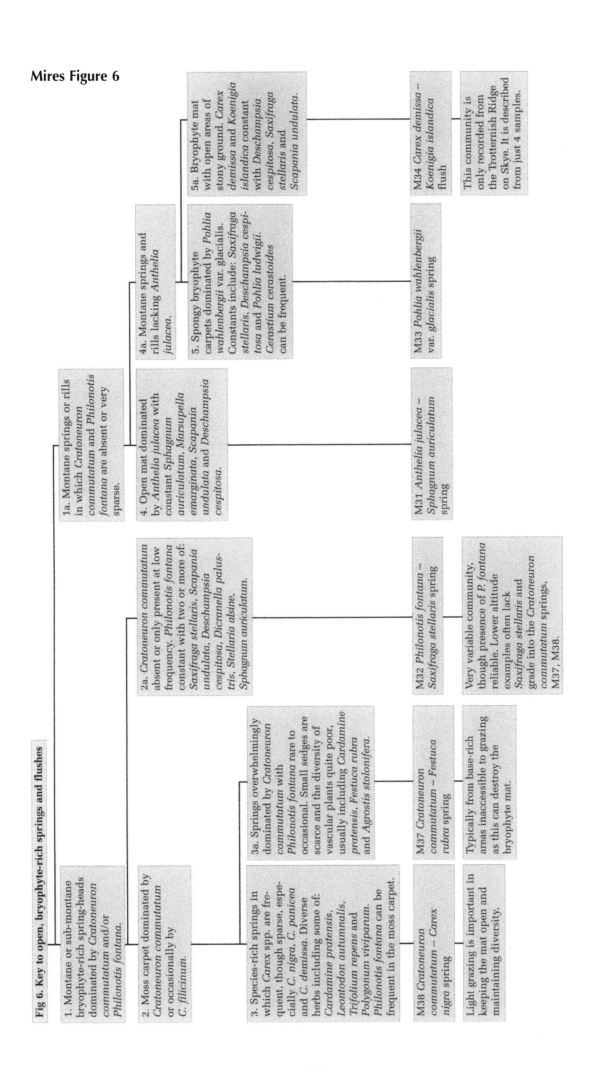

Fig 6. Key to open, bryophyte-rich springs and flushes

1. Montane or sub-montane bryophyte-rich spring-heads dominated by *Cratoneuron commutatum* and/or *Philonotis fontana*.

2. Moss carpet dominated by *Cratoneuron commutatum* or occasionally by *C. filicinum*.

3. Species-rich springs in which *Carex* spp. are frequent, though sparse, especially *C. nigra*, *C. panicea* and *C. demissa*. Diverse herbs including some of: *Cardamine pratensis*, *Leontodon autumnalis*, *Trifolium repens* and *Polygonum viviparum*. *Philonotis fontana* can be frequent in the moss carpet.

3a. Springs overwhelmingly dominated by *Cratoneuron commutatum* with *Philonotis fontana* rare to occasional. Small sedges are scarce and the diversity of vascular plants quite poor, usually including *Cardamine pratensis*, *Festuca rubra* and *Agrostis stolonifera*.

M38 *Cratoneuron commutatum – Carex nigra* spring

Light grazing is important in keeping the mat open and maintaining diversity.

M37 *Cratoneuron commutatum – Festuca rubra* spring

Typically from base-rich areas inaccessible to grazing as this can destroy the bryophyte mat.

2a. *Cratoneuron commutatum* absent or only present at low frequency. *Philonotis fontana* constant with two or more of: *Saxifraga stellaris*, *Scapania undulata*, *Deschampsia cespitosa*, *Dicranella palustris*, *Stellaria alsine*, *Sphagnum auriculatum*.

M32 *Philonotis fontana – Saxifraga stellaris* spring

Very variable community, though presence of *P. fontana* reliable. Lower altitude examples often lack *Saxifraga stellaris* and grade into the *Cratoneuron commutatum* springs, M37, M38.

1a. Montane springs or rills in which *Cratoneuron commutatum* and *Philonotis fontana* are absent or very sparse.

4. Open mat dominated by *Anthelia julacea* with constant *Sphagnum auriculatum*, *Marsupella emarginata*, *Scapania undulata* and *Deschampsia cespitosa*.

M31 *Anthelia julacea – Sphagnum auriculatum* spring

4a. Montane springs and rills lacking *Anthelia julacea*.

5. Spongy bryophyte carpets dominated by *Pohlia wahlenbergii* var. *glacialis*. Constants include: *Saxifraga stellaris*, *Deschampsia cespitosa* and *Pohlia ludwigii*. *Cerastium cerastoides* can be frequent.

M33 *Pohlia wahlenbergii* var. *glacialis* spring

5a. Bryophyte mat with open areas of stony ground. *Carex demissa* and *Koenigia islandica* constant with *Deschampsia cespitosa*, *Saxifraga stellaris* and *Scapania undulata*.

M34 *Carex demissa – Koenigia islandica* flush

This community is only recorded from the Trotternish Ridge on Skye. It is described from just 4 samples.

15

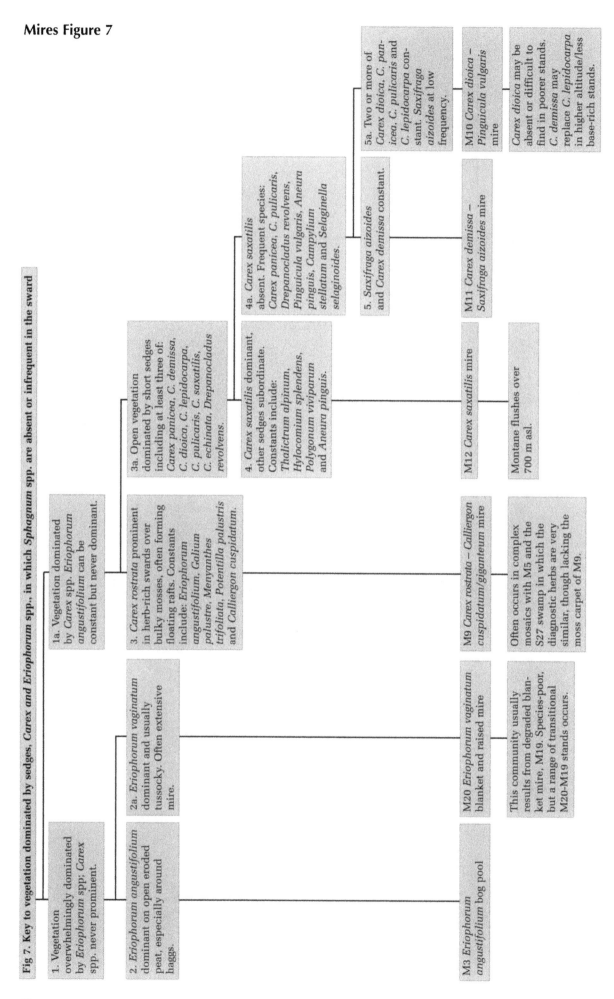

Fig 7. Key to vegetation dominated by sedges, *Carex* and *Eriophorum* spp., in which *Sphagnum* spp. are absent or infrequent in the sward

Mires Figure 8

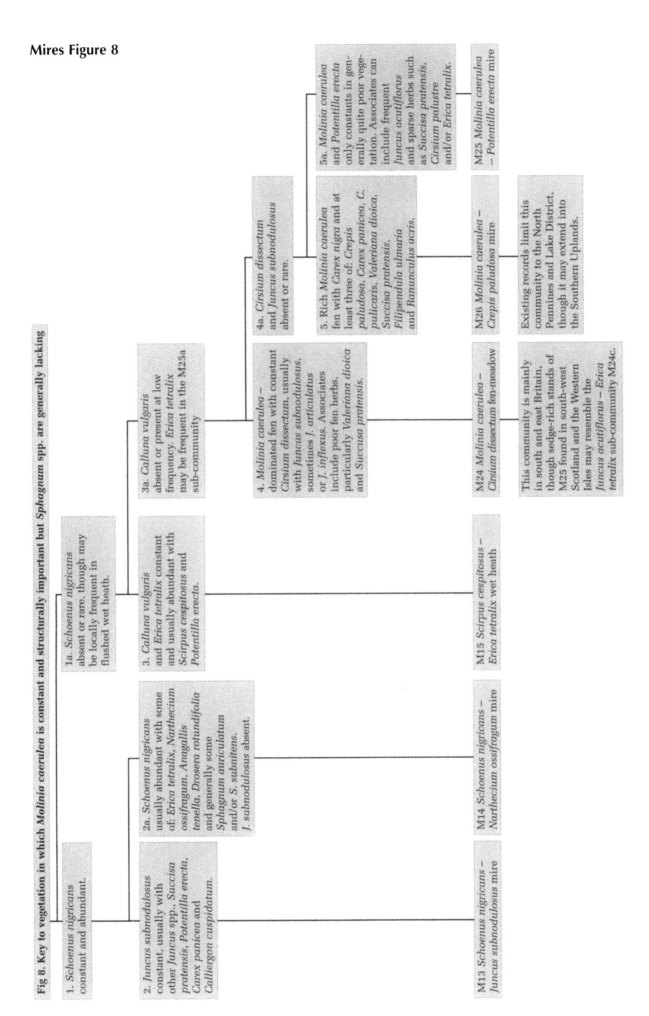

Fig 8. Key to vegetation in which *Molinia caerulea* is constant and structurally important but *Sphagnum* spp. are generally lacking

1. *Schoenus nigricans* constant and abundant.

1a. *Schoenus nigricans* absent or rare, though may be locally frequent in flushed wet heath.

2. *Juncus subnodulosus* constant, usually with other *Juncus* spp., *Succisa pratensis, Potentilla erecta, Carex panicea* and *Calliergon cuspidatum*.

2a. *Schoenus nigricans* usually abundant with some of: *Erica tetralix, Narthecium ossifragum, Anagallis tenella, Drosera rotundifolia* and generally some *Sphagnum auriculatum* and/or *S. subnitens. J. subnodulosus* absent.

3. *Calluna vulgaris* and *Erica tetralix* constant and usually abundant with *Scirpus cespitosus* and *Potentilla erecta*.

3a. *Calluna vulgaris* absent or present at low frequency. *Erica tetralix* may be frequent in the M25a sub-community

4. *Molinia caerulea* – dominated fen with constant *Cirsium dissectum*, usually with *Juncus subnodulosus*, sometimes *J. articulatus* or *J. inflexus*. Associates include poor fen herbs, particularly *Valeriana dioica* and *Succusa pratensis*.

4a. *Cirsium dissectum* and *Juncus subnodulosus* absent or rare.

5. Rich *Molinia caerulea* fen with *Carex nigra* and at least three of: *Crepis paludosa, Carex panicea, C. pulicaris, Valeriana dioica, Succisa pratensis, Filipendula ulmaria* and *Ranunculus acris.*

5a. *Molinia caerulea* and *Potentilla erecta* only constants in generally quite poor vegetation. Associates can include frequent *Juncus acutiflorus* and sparse herbs such as *Succisa pratensis, Cirsium palustre* and/or *Erica tetralix.*

M13 *Schoenus nigricans* – *Juncus subnodulosus* mire

M14 *Schoenus nigricans* – *Narthecium ossifragum* mire

M15 *Scirpus cespitosus* – *Erica tetralix* wet heath

M24 *Molinia caerulea* – *Cirsium dissectum* fen-meadow

This community is mainly in south and east Britain, though sedge-rich stands of M25 found in south-west Scotland and the Western Isles may resemble the *Juncus acutiflorus* – *Erica tetralix* sub-community M24c.

M26 *Molinia caerulea* – *Crepis paludosa* mire

Existing records limit this community to the North Pennines and Lake District, though it may extend into the Southern Uplands.

M25 *Molinia caerulea* – *Potentilla erecta* mire

Mires Figure 9

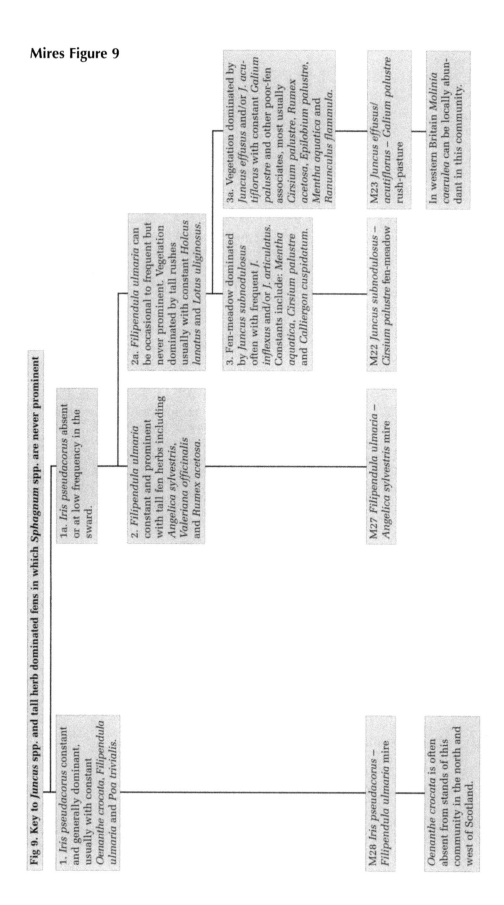

Fig 9. Key to *Juncus* spp. and tall herb dominated fens in which *Sphagnum* spp. are never prominent

1. *Iris pseudacorus* constant and generally dominant, usually with constant *Oenanthe crocata*, *Filipendula ulmaria* and *Poa trivialis*.

1a. *Iris pseudacorus* absent or at low frequency in the sward.

2. *Filipendula ulmaria* constant and prominent with tall fen herbs including *Angelica sylvestris*, *Valeriana officinalis* and *Rumex acetosa*.

2a. *Filipendula ulmaria* can be occasional to frequent but never prominent. Vegetation dominated by tall rushes usually with constant *Holcus lanatus* and *Lotus uliginosus*.

3. Fen-meadow dominated by *Juncus subnodulosus* often with frequent *J. inflexus* and/or *J. articulatus*. Constants include: *Mentha aquatica*, *Cirsium palustre* and *Calliergon cuspidatum*.

3a. Vegetation dominated by *Juncus effusus* and/or *J. acutiflorus* with constant *Galium palustre* and other poor-fen associates, most usually *Cirsium palustre*, *Rumex acetosa*, *Epilobium palustre*, *Mentha aquatica* and *Ranunculus flammula*.

M23 *Juncus effusus/ acutiflorus – Galium palustre* rush-pasture

In western Britain *Molinia caerulea* can be locally abundant in this community.

M22 *Juncus subnodulosus – Cirsium palustre* fen-meadow

M27 *Filipendula ulmaria – Angelica sylvestris* mire

M28 *Iris pseudacorus – Filipendula ulmaria* mire

Oenanthe crocata is often absent from stands of this community in the north and west of Scotland.

3 Mire community descriptions and sub-community keys

M1 *Sphagnum auriculatum* bog pool community

This bog pool community typically consists of floating masses or soft wet carpets of *Sphagnum* spp., mainly *Sphagnum auriculatum* (including var. *inundatum*) and *S. cuspidatum*, with scattered vascular plants growing on or through them or in areas of open water between. Locally, the bright orange-yellow *S. pulchrum* is conspicuous. *S. recurvum* is rare in contrast with the *Sphagnum cuspidatum/recurvum* bog pool community (M2). Other bryophytes are generally scarce, but *Cladopodiella fluitans* is characteristic at low frequencies and *Gymnocolea inflata* can also be present.

The commonest vascular plants are *Menyanthes trifoliata* and *Eriophorum angustifolium* which together make up a cover of less than 30%. In open water *Sphagnum* cover is reduced and *Utricularia* species, usually *U. minor* or locally *U. intermedia*, are sometimes present. In shallow water *Rhynchospora alba* is characteristic, and *R. fusca* is found occasionally in this community. *Narthecium ossifragum* and *Drosera* spp., particularly *D. rotundifolia*, are also occasionally present. In some areas *Carex limosa* is frequent, but is shy in flowering. Around the pool margins *Molinia caerulea* can extend down from the mire surface although its cover is generally low.

This community is confined to pools and wetter hollows on ombrogenous and topogenous mires with base-poor and oligotrophic raw peat soils in the more oceanic parts of Britain. It is a widespread component in the *Scirpus cespitosus – Eriophorum vaginatum* blanket mire (M17) in the far west of Britain including western Scotland, parts of the Lake District, Wales, and the South-West Peninsula, and the *Narthecium ossifragum – Sphagnum papillosum* valley mire (M21) in south-western valley mires with a high water table, particularly in the New Forest and Dorset.

The wetness gives some protection to this vegetation where mires are grazed or burned, but it has been reduced on many sites by draining and cutting of the peat. It has been widely lost where Erico-Sphagnion communities have been converted to Ericion heaths or their degraded derivatives. Shallow peat-digging can create flooded hollows which become suitable for recolonisation by *Sphagnum* spp., *Rhynchospora alba* and *Drosera* spp., but such locally reconstituted stands often lie in much-modified mire contexts.

No sub-communities.

M2 *Sphagnum cuspidatum/ recurvum* bog pool community

This community is typically dominated by soft wet carpets of *Sphagnum cuspidatum* or *S. recurvum*, or both. *S. pulchrum* occurs very locally, occasionally with *S. tenellum*, *S. magellanicum* or *S. papillosum*. *Sphagnum auriculatum* is rare in contrast with the *Sphagnum auriculatum* bog pool community (M1). Other bryophytes are scarce but *Polytrichum commune* or *Aulacomnium palustre* can form occasional patches and there may be scattered leafy hepatics. Vascular plants occur as scattered individuals with *Eriophorum angustifolium* and *Erica tetralix* both constant; the former often extending into deeper pools and the latter confined to drier areas. *Drosera rotundifolia* is frequent and *Narthecium ossifragum* occasional. *Andromeda polifolia*, where present, is distinctive of this vegetation type particularly around pool margins, and together with *Rhynchospora alba* it forms a clear sub-community. There may be some sedges including *Carex limosa*, *C. curta* and *C. magellanica*.

The community is typically found in pools and lawns on very wet and base-poor raw peats on ombrogenous and topogenous mires in the less oceanic parts of Britain. Its range coincides closely with that of the *Erica tetralix – Sphagnum papillosum* mire (M18) and it typically forms the pool, wet hollow and lawn elements in that community (and its degraded derivatives) on lowland raised bogs, on locally raised areas within low altitude blanket mires and in base-poor basin mires. It occurs from Wales up through the Scottish Borders and south-west Scotland with some localities in north-east Scotland.

This community has been reduced by widespread drainage and cutting of mires, so that often just small and modified fragments remain within predominantly agriculture landscapes. However this community readily colonises shallow flooded workings and appears to have expanded its coverage in sites where there has been some agricultural enrichment of the water.

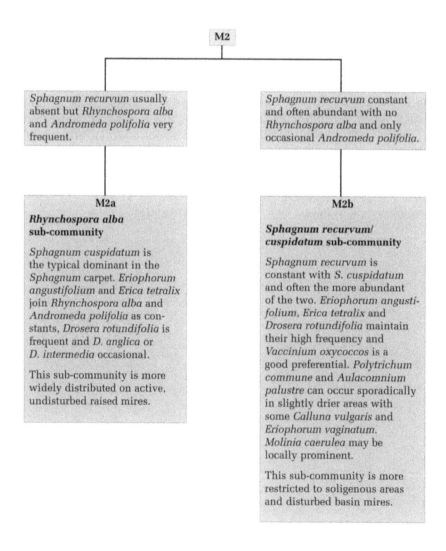

M2

Sphagnum recurvum usually absent but *Rhynchospora alba* and *Andromeda polifolia* very frequent.

Sphagnum recurvum constant and often abundant with no *Rhynchospora alba* and only occasional *Andromeda polifolia*.

M2a

Rhynchospora alba sub-community

Sphagnum cuspidatum is the typical dominant in the *Sphagnum* carpet. *Eriophorum angustifolium* and *Erica tetralix* join *Rhynchospora alba* and *Andromeda polifolia* as constants, *Drosera rotundifolia* is frequent and *D. anglica* or *D. intermedia* occasional.

This sub-community is more widely distributed on active, undisturbed raised mires.

M2b

Sphagnum recurvum/ cuspidatum sub-community

Sphagnum recurvum is constant with *S. cuspidatum* and often the more abundant of the two. *Eriophorum angustifolium*, *Erica tetralix* and *Drosera rotundifolia* maintain their high frequency and *Vaccinium oxycoccos* is a good preferential. *Polytrichum commune* and *Aulacomnium palustre* can occur sporadically in slightly drier areas with some *Calluna vulgaris* and *Eriophorum vaginatum*. *Molinia caerulea* may be locally prominent.

This sub-community is more restricted to soligenous areas and disturbed basin mires.

M3 *Eriophorum angustifolium* bog pool community

Eriophorum angustifolium is dominant here in swards where other vascular species and *Sphagnum* spp. play a relatively minor role. Its shoot density is very variable, as is the sward height which may be very short or up to half a metre or more. Usually shoots reach approximately 30 cm.

Other vascular plants attain only occasional frequency but there can be scattered small tussocks of *Eriophorum vaginatum* or *Molinia caerulea* or sparse individuals of *Drosera rotundifolia*, *Erica tetralix* or *Empetrum nigrum* ssp. *nigrum*. Bryophyte cover is also very variable and there are no constant species but *Drepanocladus fluitans* may be frequent, often growing submerged. Sparse shoots or small tufts of *Sphagnum* spp. may be present, usually *S. cuspidatum* but sometimes *S. recurvum* or *S. papillosum*.

This community is typically found as small stands on barer exposures of acid raw peat soils in depressions, erosion channels or shallow peat cuttings on a wide range of mire types. It can be found in natural hollows on surfaces of more or less intact mires but is more common among erosion features where the peat has been worn down in gullies or redistributed. It is also sometimes associated with abandoned peat workings on lowland mires. The community is particularly associated with the eroded blanket mire in the north-west of Britain, being a common feature in tracts of the *Calluna vulgaris – Eriophorum vaginatum* and *Eriophorum vaginatum* mires (M19 and M20), and it is widespread but local in lowland Erico-Sphagnion mires (M18, M21) and Ericion wet heaths (M15, M16). This community may represent a seral stage in the redevelopment of active mire vegetation following disruption.

No sub-communities.

M4 *Carex rostrata – Sphagnum recurvum* mire

This mire typically has a cover of sedges over a carpet of semi-aquatic *Sphagnum* spp. *Carex rostrata* is the commonest sedge, usually forming a rather open cover of shoots, but it can be accompanied by *C. curta, C. lasiocarpa, C. limosa* or *C. nigra* (the first two especially can be locally prominent). *Carex chordorrhiza* is a rare associate. Occasionally the taller element of the vegetation also has *Eriophorum angustifolium, Juncus effusus* or *J. acutiflorus*. There is generally an extensive wet carpet of *Sphagnum* spp. *S. recurvum* and *S. cuspidatum* are usually the most frequent and abundant species and *S. auriculatum* is also common. *Sphagnum palustre* is occasional, with sparse records for *Sphagnum subnitens* and *S. papillosum. S. squarrosum* and *S. teres* are characteristically rare, which provides a good contrast with *Carex rostrata – Sphagnum squarrosum* mire (M5). Other bryophytes are few, but *Polytrichum commune* is very frequent forming scattered patches. *Aulacomnium palustre* and *Calliergon stramineum* are very sparse.

Scattered through the ground cover are individuals of an impoverished poor-fen herb flora. The commonest species are *Agrostis canina* ssp. *canina* and *A. stolonifera* (which may be locally abundant as stoloniferous mats), *Molinia caerulea, Potentilla erecta, Galium palustre, Rumex acetosa, Viola palustris, Succisa pratensis* and *Stellaria alsine*. Usually only one or two of these are present in any one stand. *Potentilla palustris, Menyanthes trifoliata* and *Equisetum fluviatile* also may occur occasionally.

This community is characteristic of pools and seepage areas on raw peat soils of topogenous and soligenous mires where the waters are fairly acid and only slightly enriched. It can occur in bog pools on the surface of basin (and sometimes raised) mires, but is more common in obviously soligenous areas as in mire laggs and the wettest parts of water-tracks. Enrichment is slight and the pH is typically around 4. The community is of widespread but local occurrence throughout the north-west of Britain and probably remains as remnants in drained mire systems in the lowlands.

The place of this community in the terrestrialising succession is not clear and the vegetation may be very stable provided the high water table and modest irrigation are maintained. Drainage results in the demise of the more aquatic *Sphagnum* spp. and perhaps a transition to the *Carex echinata – Sphagnum recurvum/ auriculatum* mire (M6), and with grazing, may result in a spread of *Juncus* dominance.

No sub-communities.

M5 *Carex rostrata – Sphagnum squarrosum* mire

This mire is fairly heterogeneous and is characterised overall by the dominance of sedges with scattered poor-fen herbs over a patchy carpet of moderately base-tolerant *Sphagnum* spp. The commonest species throughout are *Carex rostrata* and *C. nigra*, with the former generally more extensive. *Carex lasiocarpa* can be locally prominent and *C. curta* is occasionally found. *Carex limosa* and *C. diandra* are typically absent in contrast with the *Carex rostrata – Calliergon cuspidatum/giganteum* mire (M9).

Other vascular plants are often limited to scattered individuals, but the most frequent overall are *Potentilla palustris*, *Eriophorum angustifolium*, *Menyanthes trifoliata*, *Galium palustre* and such typical poor-fen herbs as *Succisa pratensis*, *Viola palustris*, *Ranunculus flammula*, *Epilobium palustre* and *Lychnis flos-cuculi*. *Juncus effusus* can be frequent, as can *Molinia caerulea* and *Myrica gale*.

The bryophyte carpet helps define the *Carex – Sphagnum squarrosum* mire against closely related vegetation types. *Sphagnum* spp. are at least patchily prominent. Especially distinctive is the presence of *Sphagnum squarrosum* and *S. teres*. In addition *S. recurvum* and *S. palustre* are frequently encountered and *S. cuspidatum* and *S. auriculatum* are occasionally found. *Sphagnum contortum* is rare in contrast with the *Carex rostrata – Calliergon cuspidatum/giganteum* mire (M9). Other common bryophytes are *Aulacomnium palustre* and *Calliergon stramineum*.

This mire is typically found as a floating raft or on soft, spongy peats in topogenous mires and in soligenous sites with mildly acid, only moderately calcareous and rather nutrient-poor waters; the pH range is from about 4 to above 6. It is characteristically found in zonations and mosaics, the simplest being open water transitions around lakes. It can also be found around springs, seepage lines and streams where it can form part of a mixture of poor- and rich-fen communities. The community has a widespread but fairly local distribution in north-western parts of Britain. It was probably once much more widespread in the lowland south and east where relic stands may still occur.

The peat under this community is often very soft which gives the vegetation a measure of protection against the trampling and grazing effects of larger herbivores, although damage may occur during periodic dry spells. Where the community runs onto firmer peats around the margins of lakes or basins, the vegetation tends to pass to the *Carex echinata – Sphagnum recurvum/auriculatum* mire (M6). The effect of grazing on these transitions may favour the spread of *Juncus effusus*.

No sub-communities.

M6 *Carex echinata – Sphagnum recurvum/auriculatum* mire

This community has a distinct general character but includes a wide variation in composition, expressed here in four sub-communities. Essentially it is a poor-fen with small sedges or rushes dominating over a carpet of oligotrophic and base-intolerant *Sphagnum* spp. The constants are very few. Among vascular plants only *Carex echinata* has a uniformly high frequency, but *C. nigra* and *C. panicea* are common, and *C.demissa* occasional. There are two negative characters which aid definition of this community. Firstly, the general absence of more calcicolous *Carex* species, e.g. *C. dioica*, *C. pulicaris*, *C. lepidocarpa* and *C. flacca*, helps to separate this community from the Caricion davallianae rich fens (M9 to M12), and secondly, only local occurrence of species like *C. rostrata* and *C. curta* marks the vegetation off from communities like the *Carex rostrata – Sphagnum recurvum* mire (M4).

The most common vascular associates are grasses and poor-fen dicotyledons. Among the grasses, *Agrostis canina* ssp. *canina* and *Molinia caerulea* are the most common but *Anthoxanthum odoratum* is also frequent. Commonly occurring poor-fen dicotyledons include *Viola palustris* and *Potentilla erecta*, and occasionally one or more of *Galium saxatile*, *G. palustre*, *Cirsium palustre*, *Epilobium palustre*, *Succisa pratensis*, *Ranunculus flammula* or *Cardamine pratensis* may be present. Sometimes species such as *Narthecium ossifragum*, *Drosera rotundifolia* and *Erica tetralix* are found. The rushes *Juncus acutiflorus* and *J. effusus* may each be dominant in particular sub-communities.

A ground carpet of *Sphagnum* spp. is prominent and it is most frequently composed of *S. recurvum* and *S. auriculatum*, with occasional occurrence of *S. subnitens* and *S. papillosum*. There are only a few other commonly occurring bryophyte species. *Polytrichum commune* is very frequent, *Rhytidiadelphus squarrosus* is occasional and *Calliergon stramineum* and *Aulacomnium palustre* are patchy throughout. *Calliergon cuspidatum* and *Plagiothecium undulatum* are conspicuously rare.

This mire is the major soligenous community of peats and peaty gleys irrigated by rather base-poor waters in the sub-montane zone of northern and western Britain. The soils and water are quite acidic with a superficial pH usually between 4.5 and 5. It typically occurs as small stands among other mire communities, grassland and heaths and sometimes with swamp and spring vegetation. It is commonly found in tracts of unenclosed pasture on upland fringes, particularly between 200 m and 400 m (although it may be found much higher) and is ubiquitous in the upland fringes of Britain. The community is frequently grazed. This, especially where combined with drainage, can convert the community to grassland. The exclusion of herbivores would be expected to permit progress to wet scrub and woodland, although in many cases this would probably be slow and patchy.

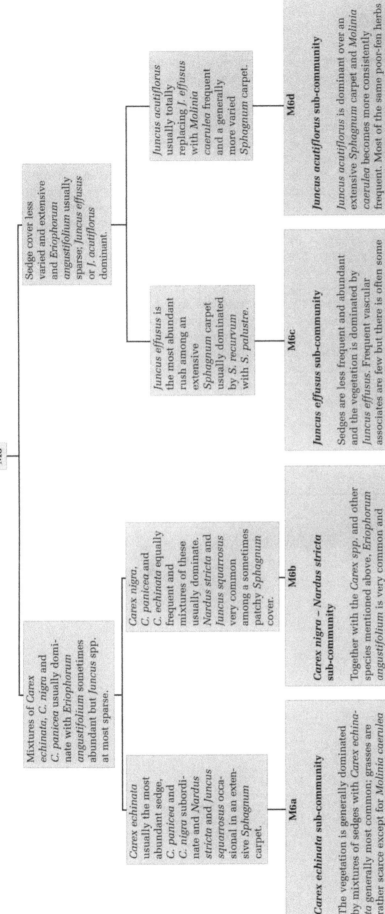

M6

Mixtures of *Carex echinata*, *C. nigra* and *C. panicea* usually dominate with *Eriophorum angustifolium* sometimes abundant but *Juncus* spp. at most sparse.

Sedge cover less varied and extensive and *Eriophorum angustifolium* usually sparse; *Juncus effusus* or *J. acutiflorus* dominant.

Carex echinata usually the most abundant sedge, *C. panicea* and *C. nigra* subordinate and *Nardus stricta* and *Juncus squarrosus* occasional in an extensive *Sphagnum* carpet.

Carex nigra, *C. panicea* and *C. echinata* equally frequent and mixtures of these usually dominate. *Nardus stricta* and *Juncus squarrosus* very common among a sometimes patchy *Sphagnum* cover.

Juncus effusus is the most abundant rush among an extensive *Sphagnum* carpet usually dominated by *S. recurvum* with *S. palustre*.

Juncus acutiflorus usually totally replacing *J. effusus* with *Molinia caerulea* frequent and a generally more varied *Sphagnum* carpet.

M6a

Carex echinata sub-community

The vegetation is generally dominated by mixtures of sedges with *Carex echinata* a generally most common; grasses are rather scarce except for *Molinia caerulea* and *Agrostis canina* ssp. *canina* which are very common. *Eriophorum angustifolium* is frequent and may dominate. Rushes are typically scarce and of low cover. The *Sphagnum* carpet is typically extensive and luxuriant. *S. palustre* is common but *S. recurvum* and *S. auriculatum* show a pattern of replacement, the latter becoming more prominent in the oceanic far west of Britain. *Drosera rotundifolia*, *Narthecium ossifragum*, *Erica tetralix*, *Juncus bulbosus/kochii* and *Menyanthes trifoliata* show some preference for the *S. auriculatum* type of flush.

This sub-community is found throughout the range of M6.

Two variants.

M6b

Carex nigra – Nardus stricta sub-community

Together with the *Carex* spp. and other species mentioned above, *Eriophorum angustifolium* is very common and *Anthoxanthum odoratum* frequent. *Juncus squarrosus* is a good preferential, but *J. effusus* is scarce and *J. acutiflorus* absent. In some stands *Sphagnum recurvum* and *S. palustre* are the commonest *Sphagnum* species with poor fen herbs such *Ranunculus flammula*, *Epilobium palustre* and *Cirsium palustre*. In contrast other stands have a greater abundance of *Sphagnum* spp. with *S. auriculatum*, *S. subnitens* and *S. papillosum* being preferential and often some *Molinia caerulea* present among the other grasses.

This sub-community is found throughout the range of M6.

Two variants.

M6c

Juncus effusus sub-community

Sedges are less frequent and abundant and the vegetation is dominated by *Juncus effusus*. Frequent vascular associates are few but there is often some *Agrostis canina* ssp. *canina*, *Potentilla erecta*, and (rather diagnostic here) *Galium saxatile*. *Carex echinata*, *Molinia caerulea* and *Viola palustris* are also fairly common. The *Sphagnum* carpet is generally extensive and luxuriant and *S. recurvum* is almost always dominant. *Polytrichum commune* remains frequent and sometimes abundant.

This sub-community is found throughout the range of M6.

Two variants.

M6d

Juncus acutiflorus sub-community

Juncus acutiflorus is dominant over an extensive *Sphagnum* carpet and *Molinia caerulea* becomes more consistently frequent. Most of the same poor-fen herbs as listed in M6c are frequent. *Sphagnum* spp. are generally abundant with *S. palustre* common throughout and *S. recurvum* or *S. auriculatum* with *S. subnitens*, *S. papillosum* and *S. capillifolium* having dominance in the carpet.

This sub-community is found throughout the range of M6.

Two variants.

M7 *Carex curta – Sphagnum russowii* mire

This mire community has prominent cyperaceous and *Sphagnum* components with a distinct northern and montane character. *Eriophorum angustifolium* and *Carex echinata* are very frequent and provide a floristic link with the *Carex echinata – Sphagnum recurvum/auriculatum* mire (M6) community which occurs at lower altitudes, but in contrast, *C. curta* is a constant often with high cover. It is often accompanied by *C. bigelowii* or *C. aquatilis* and *C. rariflora. Carex nigra* can also occur, sometimes abundantly. Larger *Juncus* spp., e.g. *Juncus effusus* and *J. acutiflorus*, are very scarce, again in contrast to M6.

The *Sphagnum* carpet is typically extensive. *Sphagnum papillosum* is common and often abundant, and *S. subnitens, S. auriculatum, S. capillifolium* or *S. recurvum* may be frequent. The high altitude species *S. russowii* is constant as is *S. lindbergii* in one of the sub-communities. The rare *S. riparium* also grows in this community. Other frequent bryophytes are *Polytrichum commune, Calliergon stramineum* or *C. sarmentosum*. Grasses play a minor role, although *Nardus stricta* is very common and *Agrostis canina* ssp. *canina* frequent. Among dicotyledons *Viola palustris* and *Galium saxatile* are most common, but are typically of low cover.

This community is confined to high altitude sites, usually above 650 m, forming small stands where peaty soils are irrigated by oligotrophic and base-poor waters. It is characteristic of hollows and drainage channels in blanket mires or flushes and seepage areas in tracts of montane moss heaths. It is an altitudinal replacement for *Carex echinata – Sphagnum recurvum/auriculatum* mire (M6) with a preponderance of montane plants. The community is mainly confined to the central Highlands of Scotland, but extends south into the Pennines and perhaps also into Wales.

Most of the occurrences of the *Carex curta – Sphagnum russowii* mire are close to or above the potential forest limit in the Scottish Highlands and the community is probably an essentially stable component of the vegetation pattern under present-day conditions.

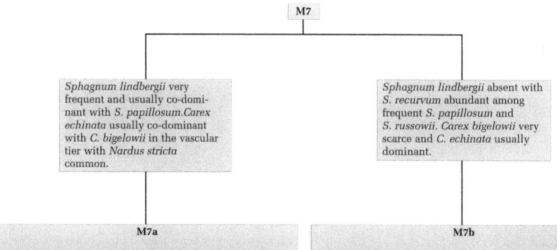

M7

Sphagnum lindbergii very frequent and usually co-dominant with *S. papillosum. Carex echinata* usually co-dominant with *C. bigelowii* in the vascular tier with *Nardus stricta* common.

Sphagnum lindbergii absent with *S. recurvum* abundant among frequent *S. papillosum* and *S. russowii. Carex bigelowii* very scarce and *C. echinata* usually dominant.

M7a

Carex bigelowii – Sphagnum lindbergii sub-community

Carex curta can be common with *C. echinata* and *C. bigelowii* but other sedges are scarce. Among the few dicotyledons *Saxifraga stellaris* is preferential. As well as the *Sphagnum* spp. mentioned above, *S. subnitens, S. auriculatum* and *S. capillifolium* are frequent and *S. recurvum* scarce. Among other bryophytes *Polytrichum commune* is frequent, but more distinctive are *Calliergon sarmentosum, Drepanocladus exannulatus, Polytrichum alpestre* and *P. alpinum.*

This sub-community is found throughout the range of M7.

M7b

Carex aquatilis – Sphagnum recurvum sub-community

Carex curta is often co-dominant with *C. echinata.* Some stands have the rare *C. aquatilis* and *C. rariflora. C. nigra* is more common in this sub-community and can be locally dominant. There is also more diversity among vascular associates with *Festuca vivipara, Agrostis stolonifera, Deschampsia flexuosa, Luzula multiflora* and *Galium saxatile* all preferential. *Nardus stricta* is only occasional. Only the *Sphagnum* spp. mentioned above are frequent. Other bryophytes are few, but *Polytrichum commune* is better represented and *Calliergon stramineum* replaces *C. sarmentosum.*

This sub-community is more local being concentrated around the Clova-Caenlochan area of the east Highlands.

M8 *Carex rostrata – Sphagnum warnstorfii* mire

This community has a dominant cover of sedges over an extensive carpet of *Sphagnum* spp. and a fairly numerous and diverse assemblage of herbs. *Carex rostrata* and *C. nigra* are the commonest sedges, the former usually more abundant and of high cover. Other poor-fen sedges, *C. panicea*, *C. echinata* and *C. demissa* occur frequently and sometimes abundantly and *C. pulicaris* is occasional.

The *Sphagnum* carpet is typically extensive, and the prominence of the base-tolerant *Sphagnum teres* and *S. warnstorfii* is quite distinctive. *Sphagnum recurvum* occurs frequently and *S. subsecundum sensu stricto*, although only occurring occasionally, is also very characteristic of this community.

Other bryophytes are numerous and in particular *Aulacomnium palustre* and *Rhizomnium pseudopunctatum* are frequent. Also distinctive are *Calliergon cuspidatum*, *C. stramineum* and less frequently *C. sarmentosum*. *Homalothecium nitens* is quite common and a good diagnostic species. *Hylocomium splendens* and *Rhytidiadelphus squarrosus* are frequently recorded.

Herbaceous associates are common but typically of low total cover. Constant species are *Epilobium palustre*, *Potentilla erecta*, *Viola palustris* and *Selaginella selaginoides*. Common grasses are *Festuca ovina* (and *F. vivipara*), *Nardus stricta*, *Anthoxanthum odoratum* and *Agrostis stolonifera*, all of which are generally present as scattered shoots or small tussocks.

This community typically occurs as small stands and is strictly confined to raw peat soils in waterlogged hollows in the montane zone of Britain where there is moderate base-enrichment by drainage from calcareous rocks. These conditions are not common. The peat deposits on which the community is found are typically quite deep, usually more than 1 m, with a high and stagnant water table. The pH of the waters and the peat is usually between 5.5 and 6. The small but distinct montane element in the flora of this community helps separate it from the *Carex rostrata – Sphagnum squarrosum* mire (M5) which is found in analogous situations in the lowlands. The community is generally confined to altitudes between 400 m and 800 m in the central Highlands except for a few examples in southern Scotland and northern England.

The frequent presence of seedlings of *Salix aurita* in stands of the community may indicate a tendency towards the development of montane willow scrub but such successions have never been seen to progress further.

No sub-communities.

M9 *Carex rostrata – Calliergon cuspidatum/giganteum* mire

This community has a diverse composition and physiognomy, even within individual stands, but is generally characterised by a fairly rich assemblage of sedges and vascular plants over a carpet of bulky mosses and localised patches of *Sphagnum* spp. The commonest large sedge is *Carex rostrata*, which is often abundant and sometimes dominant. *Carex diandra* is frequent, and *C. lasiocarpa* may accompany one or both of these species. *Carex paniculata* or *C. appropinquata* is present in some localities. Small sedges which commonly occur in this community include *C. panicea* and *C. nigra*.

Intermixed with these species, or fringing patches of them, are a variety of associates. *Potentilla palustris* and *Menyanthes trifoliata* are common and are particularly important when the vegetation forms floating rafts. Also common are *Eriophorum angustifolium*, *Equisetum fluviatile*, *E. palustre*, *Succisa pratensis*, *Pedicularis palustris*, *Cirsium palustre* and *Ranunculus flammula* often with *Galium palustre*. Less evenly distributed and usually present as scattered individuals are *Mentha aquatica*, *Caltha palustris*, *Valeriana dioica*, *Angelica sylvestris*, *Epilobium palustre* and *Lychnis flos-cuculi*. The commonest grass to occur in this community is *Molinia caerulea*, particularly in drier stands.

Bryophytes are almost always conspicuous. *Calliergon cuspidatum* is constant and *C. giganteum* and *C. cordifolium* frequent. One or more of the larger Mniaceae are also common. *Campylium stellatum* is a distinctive species of this community together with *Scorpidium scorpioides* and *Drepanocladus revolvens*.

This community is characteristic of soft, spongy peats kept permanently moist by at least moderately base-rich and calcareous waters. Waters and substrates always have a pH above 5 and usually above 6. It is commonest in wetter parts of topogenous mires in hollows or old peat-workings, but also around springs, laggs of raised mires and mowing marshes. The community is widespread but local, ranging from south-east England, particularly in Broadland, to Wales and northern England and through most of Scotland.

The community is limited by the fairly sparse occurrence of suitable natural situations and in the lowland south-east by wetland drainage and the cessation of shallow peat-digging. It is typically too wet to be grazed but in some areas it occurs within mowing marsh that is periodically cropped. Throughout its range, at least at the lower altitudes and in less remote sites, the *Carex rostrata – Calliergon cuspidatum/giganteum* mire is probably a successional stage to *Salix pentandra – Carex rostrata* woodland (W3) in the north and *Salix cinerea – Betula pubescens – Phragmites australis* woodland (W2) in the south-east, although development to woodland may be hindered by a high water table or by the periodic mowing of the vegetation. There is also the possibility that under certain conditions this mire type is seral to the development of poor-fen and ombrogenous mire through the local formation of *Sphagnum* nuclei.

M9

Carex rostrata usually dominant, sometimes with *C. lasiocarpa* or more locally *C. diandra* or *Schoenus nigricans*. *Calliergon cuspidatum* common but other *Calliergon* species and larger Mniaceae at most local. Ground carpet usually dominated by mixtures of *Campylium stellatum, Scorpidium scorpioides* and *Drepanocladus revolvens*.

Carex rostrata and *C. diandra* constant, either or both dominant with or without *C. lasiocarpa*. Herbaceous associates often numerous and lush. *Campylium stellatum* only occasional and *Scorpidium scorpioides* and *Drepanocladus revolvens* rare, but *Calliergon cordifolium* and *C. giganteum* common, often with large Mniaceae.

M9a

Campylium stellatum – Scorpidium scorpioides sub-community

Smaller sedges are more numerous than in M9b; *Carex panicea* and *C. nigra* retain their high frequency and *C. limosa* and *C. echinata* are strongly preferential. Herbaceous associates are variable and not very rich, and total cover is low, giving an open community. *Menyanthes trifoliata* and *Potentilla palustris* can be prominent, but most species occur as scattered plants. Bryophytes are usually prominent as indicated above. Clumps of *Sphagnum* spp. are occasional, notably so for the base-tolerant *Sphagnum contortum*.

This sub-community is largely north-western in range.

M9b

Carex diandra – Calliergon giganteum sub-community

The pattern of dominance is more variable here as indicated above. *Juncus subnodulosus* is locally abundant in eastern England. Herbaceous associates are more numerous: *Potentilla palustris, Menyanthes trifoliata* and *Filipendula ulmaria* can be prominent with *Angelica sylvestris, Epilobium palustre, Lychnis flos-cuculi, Valeriana dioica, Caltha palustris, Cardamine pratensis* and *Mentha aquatica* all frequent. Bryophytes are often extensive with *Calliergon giganteum, C. cordifolium, Plagiomnium rostratum* and *P. affine* showing their maximum development in this sub-community.

This sub-community occurs throughout the range mostly in topogenous mires.

M10 *Carex dioica – Pinguicula vulgaris* mire

The community includes a range of distinctive calcicolous flush vegetation in which the bulk of the sward is composed of small sedges, dicotyledons and bryophytes. There are marked variations in proportions of frequent species, as indicated in the three sub-communities and their variants. Essentially this is a small sedge mire with *Carex dioica*, *C. hostiana*, *C. lepidocarpa*, *C. panicea* and *C. pulicaris* as constants which are also often abundant. *Carex nigra* is frequent and *C. echinata* and *C. flacca* common. Other prominent Cyperaceae are *Eriophorum angustifolium*, a constant, and *Eleocharis quinqueflora*.

Some rushes and grasses occur frequently. *Juncus articulatus* is a constant and *J. bulbosus/kochii* is typical of less base-rich sites. *Molinia caerulea* is the commonest grass with *Festuca ovina* more variable in occurrence. Other herbs generally occur as scattered plants. The frequent occurrences of *Pinguicula vulgaris* and *Selaginella selaginoides* are very distinctive of this community. *Potentilla erecta* and *Succisa pratensis* are also common species in this community but *Equisetum palustre* and *Euphrasia officinalis* agg. are more variable in occurrence. Other species are characteristic of particular sub-communities.

Bryophytes are always obvious, often comprising 50% or more of the ground cover. Calcicolous species such as *Campylium stellatum*, *Aneura pinguis*, *Drepanocladus revolvens*, *Ctenidium molluscum*, *Fissidens adianthoides* and *Cratoneuron commutatum* are frequent, together with *Bryum pseudotriquetrum*. Such assemblages of calcicolous species provide a strong definition for the community against its counterparts in the Caricion nigrae (M5-M7) and in flushed Oxycocco – Sphagnetea mires (M14-M21) which occupy similar but more base-poor situations.

This community is typically a soligenous mire of mineral soils and shallow peats kept very wet by base-rich, calcareous and oligotrophic waters. The pH of flushing waters is high, usually between 5.5 and 7.0 or sometimes higher, and the composition of this community is one of the most calcicolous of British mires. It is found in small stands, often associated with spring and rill vegetation, within grasslands and more occasionally in ombrogenous mires and around topogenous mires. Typically the *in situ* formation of peat is limited, a feature which helps to distinguish the habitat of the community from that of base-rich basins where it is replaced by vegetation like the *Carex rostrata – Calliergon cuspidatum/giganteum* mire (M9) and the *Carex rostrata – Sphagnum warnstorfii* mire (M8). It is predominantly a community of north-west Britain from Wales and the Pennines northwards, developed in the cool, wet climate.

The community typically occurs in unenclosed uplands and most of the stands are grazed and trampled by large herbivores. It is probably these factors, combined with nutrient impoverishment and the often strong and scouring effect of the irrigation, which play a major part in maintaining the community in its generally rich, varied and open state. Most stands would probably progress to Alno – Ulmion scrub or woodland (W7, W9) if grazing were withdrawn. However, at higher altitudes the vegetation may be a climatic climax.

M10

Carex hostiana, C. pulicaris and *C. nigra* all frequent with *Eriophorum angustifolium* and *Molinia caerulea. Potentilla erecta* and *Succisa pratensis* common with *Ctenidium molluscum* and *Fissidens adianthoides* frequent.

Species listed opposite scarce or absent. *Gymnostomum recurvirostrum* or less commonly *Catascopium nigritum* forming hummocks with *Drepanocladus revolvens* and *Cratoneuron commutatum* often abundant.

Carex demissa and *C. echinata* occasional to frequent; *C. lepido-carpa* and *C. flacca* scarce and *C. pulicaris* patchy. *Juncus bulbo-sus* and *Erica tetralix* quite common.

Carex demissa and *C. echinata* scarce, but *C. lepidocarpa, C. hostiana, C. pulicaris* and *C. flacca* frequent. *Juncus bulbosus* and *Erica tetralix* only of local significance.

M10c

Gymnostomum recurvirostrum sub-community

Vascular plants have low individual and total cover; much more prominent are the conspicuous moss hummocks, particularly of *Gymnostomum recurvirostrum* which are up to 30 cm high and 60 cm across. There is much bare ground with a fragmentary cover of vascular plants; good preferentials are *Plantago maritima, Sagina nodosa* and *Minuartia verna. M. stricta* is restricted to this sub-community in its only British locus.

This striking vegetation is only recorded from Upper Teesdale.

M10a

Carex demissa – Juncus bulbosus/kochii sub-community

This comprises the less calcicolous types of M10 with vascular plants predominant. *Carex panicea, C. dioica, C. hostiana* and *C. nigra* remain very frequent with the species listed above. More calcicolous herbs such as *Briza media, Primula farinosa, Linum catharticum* and *Sesleria albicans* are usually poorly represented. Bryophytes are generally less prominent in the turf; *Bryum pseudotriquetrum, Fissidens adianthoides* and *Ctenidium molluscum* are rather uncommon and *Campylium stellatum* and *Scorpidium scorpioides* rather more prominent.

This sub-community is largely restricted to Scotland and the Lake District with outlying occurrences in north-west Wales and Upper Teesdale.

Three variants.

M10b

Briza media – Primula farinosa sub-community

Vascular plants are also prominent but many swards are open with extensive bare ground. Calcicoles and more mesophytic herbs well represented. *Carex lepidocarpa, C. hostiana* and *C. pulicaris* are consistently frequent and commonly accompanied by *C. flacca*. Among the preferentials *Briza media, Primula farinosa, Linum catharticum, Sesleria albicans* and *Equisetum variegatum* are frequent. *Juncus bulbosus/ kochii, Erica tetralix, Narthecium ossifragum* and *Drosera rotundifolia* are reduced in their occurrence. Among the bryophytes *Aneura pinguis, Ctenidium molluscum* and *Fissidens adianthoides* are consistently frequent.

This sub-community is predominantly found in northern England.

Three variants.

M11 *Carex demissa – Saxifraga aizoides* mire

This vegetation is typically open with rich mixtures of small sedges, other herbs and bryophytes among water-scoured runnels and with much exposed silt and rock debris. There is a strong floristic link with *Carex dioica – Pinguicula vulgaris* mire (M10) but the Arctic-Alpine element of the vegetation is much more pronounced in this community than it is in M10. Typically there is no single dominant. *Carex demissa, C. panicea* and *C. pulicaris* are very frequent throughout and *C. flacca* and *C. dioica* are common in some variants. *Juncus articulatus* is a constant, *Eriophorum angustifolium* is frequent as is *Eleocharis quinqueflora* at lower altitudes. At higher altitudes *Juncus triglumis* is constant and *Tofieldia pusilla* becomes frequent. By comparison with M10, *C. lepidocarpa* and *C. hostiana* are much less common and *C. nigra* and *C. echinata* also rather scarce. At higher altitudes there is an Arctic-Alpine element with *Juncus triglumis* being constant and *Tofieldia pusilla* becoming more frequent. Grasses are typically low in cover. *Festuca ovina/vivipara* is common and *Agrostis stolonifera* occasional. At higher altitudes *Deschampsia cespitosa* (including *D. alpina*), *Nardus stricta*, *Anthoxanthum odoratum*, *Agrostis canina* ssp. *canina* and *Festuca rubra* may be present.

Other herbs found in this community include *Pinguicula vulgaris* and *Saxifraga aizoides* which are both constant and *Selaginella selaginoides* which is very frequent. The montane element of this community includes *Thalictrum alpinum*, which is very common at higher altitudes, and occasionally *Saxifraga stellaris, S. oppositifolia* and *Alchemilla filicaulis* ssp. *filicaulis*. Typically all these species occur in an uneven and broken sward.

Bryophytes are frequent and varied. *Aneura pinguis, Campylium stellatum, Drepanocladus revolvens, Bryum pseudotriquetrum*, and at lower altitudes, *Cratoneuron commutatum, Fissidens adianthoides, Ctenidium molluscum* and *Scorpidium scorpioides* are all common. The montane moss *Blindia acuta* can be prominent at higher altitudes and is a good preferential for this community.

This community is characteristic of open, stony flushes, strongly irrigated with moderately base-rich waters, on generally steep slopes in sub-montane and montane parts of Britain. Although the community can occur almost at sea level in the far north-west of Scotland, it is generally confined to high altitudes. It is always associated with calcareous bedrocks having a soil pH range of 5.5 to 7.0. Flushing is vigorous and erosion of the surface is therefore often pronounced and the soil cover little more than scoured accumulations of silt and organic matter with exposed rock debris. It is largely confined to Scotland, but also present in the Lake District, and more locally in the Southern Uplands, the northern Pennines and north Wales.

The community is normally grazed and this grazing may help maintain the open structure and help prevent the development of a woody cover. However, colonisation of trees and scrubs would be slow due to the climatic conditions in which the community occurs and at higher altitudes the mire is probably a climatic climax.

```
                            M11
      ┌──────────────────────┴──────────────────────┐
```

Juncus triglumis and *Thalictrum alpinum* constant, but *Eleocharis quinqueflora* uncommon. *Deschampsia cespitosa, Nardus stricta* and *Anthoxanthum odoratum* frequent and *Alchemilla alpina* occasional. *Cratoneuron commutatum, Scorpidium scorpioides* and *Fissidens adianthoides* all scarce.

Eleocharis quinqueflora constant with *Juncus triglumis* and *Thalictrum alpinum* becoming more frequent at higher altitudes. The grasses found in M11a at most occasional. *Cratoneuron commutatum* and *Scorpidium scorpioides* very common and abundant, often with *Fissidens adianthoides*.

M11a
Thalictrum alpinum – Juncus triglumis sub-community

There is an obvious montane element in the vegetation as indicated in the species above. *Saxifraga aizoides, Carex demissa* and *C. panicea* are usually the most abundant vascular plants with *Blindia acuta, Campylium stellatum* or *Drepanocladus revolvens* predominating among the mosses.

This is the typical form of M11 at higher altitudes and is virtually confined to Scotland.

Two variants.

M11b
Cratoneuron commutatum – Eleocharis quinqueflora sub-community

In this sub-community M11 grades into M10 with more extreme montane plants, except *Saxifraga aizoides* and *Blindia acuta*, much more poorly represented; and in more southerly stands even these become rare. *Eleocharis quinqueflora* is constant and sometimes abundant, rivalling the sedges, among which *Carex hostiana* and, in wetter stands, *C. rostrata* are sometimes found. Vascular plant cover typically more extensive than in M11a.

This sub-community is also frequent in Scotland at lower altitudes and in most of the English and Welsh stands.

M12 *Carex saxatilis* mire

Carex saxatilis is typically dominant in this montane mire with a distinctive assemblage of associates. The sward is generally less than 20 cm high and rather open with patches of soil. *Carex demissa*, *C. echinata* and *C. nigra* are very frequent and can be abundant. *Carex bigelowii* is fairly consistent, especially in grassy transitions to surrounding swards.

Eriophorum angustifolium is also frequent attaining a cover of more than 10%. Almost all other herbs occur as scattered individuals. *Selaginella selaginoides* and *Pinguicula vulgaris* are both very common as in other calcicolous flushes, but more distinctive are *Thalictrum alpinum*, *Polygonum viviparum*, and *Juncus triglumis*. *Saxifraga aizoides* is infrequent, in contrast to *Carex demissa – Saxifraga aizoides* mire (M11). Also common are the poor-fen herbs *Viola palustris*, *Caltha palustris* and *Agrostis canina* ssp. *canina*.

Bryophytes are an important element of the vegetation although apart from the constant *Drepanocladus revolvens*, cover of individual species is low. *Aneura pinguis* is frequent and *Bryum pseudotriquetrum*, *Blindia acuta*, *Campylium stellatum* and *Calliergon trifarium* are occasional. *Hylocomium splendens* is also a constant as is *Scapania undulata*. There can also be some small patches of *Sphagnum* spp.

This mire type is strictly confined to margins of high-montane flushes irrigated with base-rich and calcareous waters perhaps influenced by long snow-lie. It typically occurs as small stands bordering rills or more strongly irrigated soligenous mires. The soils that this community is found on, though continuously irrigated, are not of especially high pH, ranging from 4.6 to 6.3. The community is fairly widespread but local on peaks above 750 m through the southern and central Scottish Highlands with scattered localities in north-west Scotland.

The physical effects of flushing, snow-melt, cryoturbation, and solifluctional flow result in the continual instability of the substrate on which this community is found and this is important in maintaining open stony areas where rare Arctic-Alpine sedges and rushes find a niche. It is possible that grazing prevents colonisation by Arctic-Alpine willows; however, in the extreme environment in which it occurs the community is probably a climatic climax.

No sub-communities.

M13 *Schoenus nigricans – Juncus subnodulosus* mire

In this community *Schoenus nigricans* is typically very frequent and consistently associated with other distinctive floristic features. It is generally dominant (although it may be absent from fragmentary stands) giving a grey-green appearance to the vegetation. Commonly it is intermixed with *Juncus subnodulosus*, and where this predominates the vegetation is olive-green in spring and reddish brown in winter. *Molinia caerulea* is also constant. These species form a rough sward about 50 cm in height with smaller herbs growing inbetween. Sedges are often important, particularly *Carex panicea*, *C. lepidocarpa* and *C. flacca*. Where the summer water table is close to the surface, species such as *Equisetum palustre*, *Pedicularis palustris*, *Mentha aquatica*, *Valeriana dioica* and *Cardamine pratensis* occur, sometimes with *Parnassia palustris*, *Pinguicula vulgaris* and *Eriophorum latifolium*. A variety of orchids are found, particularly *Epipactis palustris*. Taller herbs can be locally abundant with *Succisa pratensis* being most common. *Phragmites australis* is also frequent, particularly in ungrazed stands. On drier areas and particularly tops of *Schoenus* tussocks less calcicolous plants are found, most frequently the constant *Potentilla erecta* and *Erica tetralix*.

Bryophytes vary in cover and species but can be very extensive. The commonest throughout are *Campylium stellatum* and *Calliergon cuspidatum*. Other frequent species include *Drepanocladus revolvens*, *Aneura pinguis*, *Cratoneuron commutatum* and *C. filicinum*.

This community is confined to peat or mineral soils, in and around lowland mires irrigated by base-rich, highly calcareous, and oligotrophic waters. It is often found below springs and seepage lines and on flushed margins of valley mires, but also extends into topogenous basins provided there is close contact with waters draining from lime-rich substrates. The flushing waters typically have pH between 6.5 and 8. It is widespread but local throughout lowland England and Wales, but is restricted by natural scarcity of suitable habitat and its extensive destruction.

The structure and floristics of this community are often influenced by grazing and some stands have been affected by mowing and burning. Shallow peat-digging has been locally important in providing a suitable habitat for the community but more drastic treatment of mires, particularly draining and eutrophication, have reduced its extent and eliminated it from some areas.

M13

General floristic and structural features well preserved, with at least some of *Anagallis tenella, Pedicularis palustris, Angelica sylvestris, Cirsium palustre, Mentha aquatica, Equisetum palustre* and *Phragmites australis* frequent.

Juncus subnodulosus and *Molinia caerulea* often very abundant with *Schoenus nigricans* markedly reduced in vigour. *Festuca rubra, Holcus lanatus, Agrostis canina* and *A. stolonifera* frequent and tall herbs, orchids and bryophytes patchy.

Carex hostiana and *C. pulicaris* very frequent among an abundant and diverse small herb flora in runnels with *Briza media, Pinguicula vulgaris, Linum catharticum,* and *Juncus articulatus* common.

Many smaller runnel herbs sporadic, but *Caltha palustris* and *Valeriana dioica* become common and taller dicotyledons are often prominent with *Filipendula ulmaria, Eupatorium cannabinum* and *Lychnis flos-cuculi* frequent.

M13a
Festuca rubra – Juncus acutiflorus sub-community

This comprises the more impoverished stands of M13. Apart from the reduction or even absence of *Schoenus nigricans* and presence of the species mentioned above, *Carex panicea, C. lepidocarpa,* and *C. flacca* are also important in runnels. *Anagallis tenella* is totally absent and *Pedicularis palustris, Epipactis palustris* and other orchids are very scarce. The commonest herbs are *Succisa pratensis* and *Hydrocotyle vulgaris.* Bryophytes are generally sparse and low in number with *Calliergon cuspidatum* the commonest species.

This sub-community occurs through the range of M13.

M13b
Briza media – Pinguicula vulgaris sub-community

This kind of Schoenetum is strikingly rich. Mixtures of *Schoenus nigricans, Juncus subnodulosus* and *Molinia caerulea* usually share dominance but the small herbs of runnels are especially distinctive. Apart from the species mentioned above, *Parnassia palustris* is frequent often with mixtures of orchids including *Gymnadenia conopsea* var. *densiflora, Dactylorhiza fuchsii, D. majalis* ssp. *purpurella* and *Epipactis palustris.* Along with *Succisa pratensis* and *Serratula tinctoria,* taller herbs are represented by frequent *Angelica sylvestris, Cirsium palustre, Eupatorium cannabinum* and *Oenanthe lachenalii.* Bryophytes are quite numerous and sometimes of high cover.

This sub-community occurs in Anglesey and East Anglia.

M13c
Caltha palustris – Galium uliginosum sub-community

Schoenus nigricans, Juncus subnodulosus and *Molinia caerulea* remain of structural importance but are variously augmented by *Carex rostrata, C. diandra, C. elata, Cladium mariscus* and sometimes *Phragmites australis.* Runnels are well developed but smaller preferentials of M13b are only occasional. The commonest species are *Carex panicea, C. lepidocarpa, Mentha aquatica, Hydrocotyle vulgaris* together with the preferentials *Caltha palustris* and *Valeriana dioica.* In addition to *Epipactis palustris* there is often *Dactylorhiza incarnata, D. majalis* ssp. *praetermissa* and sometimes *D. traunsteineri.* Taller herbs are common, as listed above, with sprawling *Galium uliginosum* and less commonly *G. palustre.* A pool element is sometimes present, with *Carex rostrata* and *C. diandra* together with *Menyanthes trifoliata, Equisetum fluviatile* and *Utricularia* species.

This sub-community is concentrated in East Anglia.

M14 *Schoenus nigricans – Narthecium ossifragum* mire

This mire type includes mildly calcicolous *Schoenus* vegetation of south-west England that is not readily integrated into *Schoenus nigricans – Juncus subnodulosus* mire (M13) and with a less varied flora. *Schoenus nigricans* is usually dominant and *Molinia caerulea* is generally abundant. A mixture of these two species usually cover the ground. *Juncus subnodulosus* is absent in contrast to M13. Small calcicolous herbs are generally absent. *Narthecium ossifragum* and *Anagallis tenella* are constants while *Drosera rotundifolia*, growing on cushions of *Sphagnum*, is less common. *Erica tetralix*, or occasionally *Calluna vulgaris*, grows on *Schoenus* or *Molinia* tussocks. Some stands have a local abundance of *Myrica gale*.

Bryophytes are variable and also less calcicolous in character than in M13. *Campylium stellatum* and *Aneura pinguis* are frequent and together with *Scorpidium scorpioides* and, less commonly, *Drepanocladus revolvens*, can form extensive mats in runnels. *Sphagnum* spp. are a consistent feature, particularly on tussocks. *Sphagnum subnitens* is most common and *S. auriculatum* is frequent. *Hypnum jutlandicum* is preferential and there may be patches of hepatics including *Kurzia pauciflora* and *Calypogeia* species.

This community is characteristic of peats and mineral soils irrigated by moderately base-rich and calcareous ground waters of a pH range between 5 and 7. It characteristically occurs as isolated flushes among wet heath and moorland vegetation, but it is also associated with soligenous zones within valley mires. The community occurs very locally in Cornwall, east Devon, south-east Dorset and the New Forest. It may also be found in Wales but it is replaced in comparable situations on north-western blanket bogs by *Schoenus*-dominated stands of *Scirpus cespitosus – Erica tetralix* wet heath (M15).

The community only occurs very locally. This is partly because of the natural scarcity of suitable habitats, but also because of the reduction in its extent by human activities such as drainage and agricultural improvement. Occasional burning and light grazing are also of common occurrence over the tracts of heath in which this kind of mire usually occurs, although these activities are probably not very damaging. In the absence of grazing or burning it is expected that some stands of this community would progress towards wet woodland.

No sub-communities.

M15 *Scirpus cespitosus – Erica tetralix* wet heath

This is a vegetation type with few constants and a wide variation in the pattern of dominance and in associated flora. *Molinia caerulea*, *Scirpus cespitosus*, *Erica tetralix* and *Calluna vulgaris* are all of high frequency and it is mixtures of these species that give the vegetation its general character. However, sometimes one or two of them may be missing and their relative proportions are very diverse. *Molinia* is the most consistent overall and often abundant; in other stands *Scirpus* is very prominent and both may share dominance with *Calluna*. *Molinia* may also dominate with *Scirpus* or with *Erica tetralix*. The shrubby species *Erica cinerea*, *Vaccinium myrtillus* and *Myrica gale* are important in particular sub-communities. Other common species are *Potentilla erecta*, and in moister stands, *Polygala serpyllifolia*, *Narthecium ossifragum* and *Eriophorum angustifolium*. By contrast *E. vaginatum* is notably scarce.

There are few bryophytes common throughout. There are usually some *Sphagnum* spp. but they do not form the luxuriant carpets of the Sphagnetalia mires (M17-M21). The most frequent species overall are *Sphagnum capillifolium* and *S. subnitens*. *Sphagnum palustre*, *S. recurvum* and *S. auriculatum* can become common in wetter stands. Lichens do not appear consistently but *Cladonia* spp. can be locally prominent.

This wet heath community is characteristic of moist and generally acid and oligotrophic peats and peaty mineral soils in the wetter western and northern parts of Britain. It is associated with thinner or better drained areas of ombrogenous peat with a surface pH typically between 4 and 5. The community is particularly well represented in the west and south-west of Scotland, through Wales and less extensively in the Lake District, Dartmoor and Exmoor.

Grazing and burning have important effects on the floristics and structure of this community, and draining and peat-cutting have extended its coverage to formerly deeper and wetter peats. Without burning or grazing, less damaged stands may be able to revert to blanket mire. However, cessation of burning, especially on peat that is well aerated or where there has been drainage, may precipitate a vigorous expansion of *Molinia*. Although progression to woodland is theoretically possible over most, if not all, of its altitudinal range, widespread deforestation has often removed potential seed-parents, and continued grazing by livestock and deer and sporadic burning may be enough to set back succession continually. However, extensive tracts of this kind of vegetation have been replaced by coniferous forest after the ground has been drained.

M15

Narthecium ossifragum fairly common, but other species listed opposite all scarce.

Narthecium ossifragum and Eriophorum angustifolium frequent with Sphagnum palustre common. Myrica gale often found.

Narthecium ossifragum fairly common, but other species listed opposite all scarce.

Usually small stands, often in soakways. Sphagnum carpet extensive with frequent Sphagnum recurvum and S. subnitens. Drosera rotundifolia common and scattered small sedges such as Carex echinata, C. panicea and C. nigra.

Drosera rotundifolia and small sedges at most occasional. Sphagnum carpet patchy, but S. papillosum quite frequent.

Erica cinerea frequent, sometimes abundant, but Vaccinium myrtillus rare. Racomitrium lanuginosum common and Cladonia spp. often abundant.

Erica cinerea rare but Vaccinium myrtillus frequent, commonly with Nardus stricta, Juncus squarrosus and Deschampsia flexuosa. Racomitrium lanuginosum and Cladonia spp. scarce.

M15a

Carex panicea sub-community

This is the richest and most floristically distinct sub-community. Molinia caerulea and Erica tetralix retain high frequency; Scirpus cespitosus and Calluna vulgaris are more sparse. Myrica gale sometimes has local abundance, but E. cinerea and Vaccinium myrtillus almost totally absent. Potentilla erecta and Polygala serpyllifolia are very commonly found with Narthecium ossifragum and Erica angustifolium. Drosera rotundifolia is preferential with a variety of species such as Carex panicea, C. echinata, C. nigra, C. pulicaris, C. demissa, C. dioica, Selaginella selaginoides, Pinguicula vulgaris, Succisa pratensis, Viola palustris, Juncus bulbosus and Dactylorhiza maculata ssp. maculata. The Sphagnum carpet is also distinctive, as indicated above, with Sphagnum palustre also often abundant and with S. capillifolium patchy.

This and the Typical sub-community are particularly common in the west of Scotland.

M15b

Typical sub-community

The dominants here are very variable. Scirpus cespitosus and Calluna vulgaris may share dominance, or Calluna and Molinia caerulea may predominate. Molinia and Erica tetralix or Molinia alone may be dominant. Myrica gale is quite common but not abundant. Narthecium ossifragum and Eriophorum angustifolium are frequent as in M15a, but small sedges are generally sparse, with only Carex panicea and C. echinata occasional and fen associates very uncommon. Nardus stricta and Juncus squarrosus may show local prominence. Eriophorum vaginatum is a low-cover occasional. Sphagnum papillosum is frequent and locally abundant and Odontoschisma sphagni often present. In some stands Sphagnum spp. are sparse and mosses such as Racomitrium lanuginosum, Dicranum scoparium, Hypnum cupressiforme and Campylopus paradoxus provide most of the cover.

This and the Carex panicea sub-community in the west of Scotland.

M15c

Cladonia spp. sub-community

All four possible dominants have high frequency but Calluna vulgaris usually predominates. Potentilla erecta remains constant, but Polygala serpyllifolia and Narthecium ossifragum are less common and Eriophorum angustifolium and Myrica gale very scarce. Sphagnum spp. are only poorly represented and Hypnum cupressiforme/jutlandicum and Racomitrium lanuginosum become frequent. Cladonia spp. are abundant, particularly Cladonia impexa and C. uncialis together with C. arbuscula, C. pyxidata, C. coccifera and C. gracilis.

This sub-community is especially common in the drier regions of the distribution of M15.

M15d

Vaccinium myrtillus sub-community

Mixtures of Molinia caerulea and Calluna vulgaris generally dominate with Scirpus cespitosus and Erica tetralix both rather uneven, often with some Vaccinium myrtillus. Commonly there are small tussocks of Nardus stricta, Juncus squarrosus, Deschampsia flexuosa and more occasionally some Anthoxanthum odoratum, Festuca ovina/vivipara, F. rubra, Luzula multiflora and Carex pilulifera. Sphagnum spp. are infrequent, their place being taken by Hypnum cupressiforme/jutlandicum, Dicranum scoparium, Pleurozium schreberi, Plagiothecium undulatum, Polytrichum commune and Rhytidiadelphus loreus. Racomitrium lanuginosum and Cladonia spp. are scarce.

This sub-community is especially common in the drier regions of the distribution of M15.

M16 *Erica tetralix – Sphagnum compactum* wet heath

This community is characteristically dominated by mixtures of *Erica tetralix*, *Calluna vulgaris* and *Molinia caerulea*, but their proportions are very variable, being influenced by differences in the water regime and trophic state of the soils, and also by grazing and burning. *Erica tetralix* is often vigorous, particularly on wetter soils, while *Calluna* is often subordinate and weak (although it may be abundant in drier stands or where controlled burning is carried out). No other sub-shrubs attain a high frequency, although *Erica cinerea* and *Ulex gallii* may be abundant in transitions to drier heaths in south-west England and *E. cinerea* and *U. minor* can occur in similar situations further east. In some situations *Molinia* may be dominant.

This community may have no other, or only sporadic, vascular associates. The commonest vascular associate, where present, is *Scirpus cespitosus. Eriophorum angustifolium* and *Narthecium ossifragum* are quite frequent, as is *Drosera rotundifolia* in wetter hollows. *Myrica gale* occurs occasionally, sometimes with local abundance.

Most characteristic of the bryophyte layer in drier situations is *Sphagnum compactum*, a constant and strong preferential for the community. In wetter places *S. tenellum* may be present. These species may occur as scattered cushions or form a continuous carpet, sometimes with several other *Sphagnum* spp. as well as a number of other bryophytes, between the dominants. Lichens may also be present, especially larger *Cladonia* species such as *C. impexa* and *C. uncialis*.

This wet heath community is found on acid and oligotrophic mineral soils or shallow peats that generally have a surface pH of between 3.5 and 4.5 and that are at least seasonally water-logged. It is characteristic of the south of lowland England, being particularly associated with the surrounds of valley mires maintained by a locally high water table. It is also found through Wales, and in northern England and Scotland, where it extends on to thin ombrogenous peats at higher altitudes.

Grazing and burning are important in maintaining the vegetation, and burning is able to transform the appearance of particular stands over short periods of time, producing considerable structural diversity within a small area. Without any grazing or burning most stands would probably progress to woodland, and this has happened to some stands lying within tracts of heath on commons in south-east England where traditional management has fallen into disuse. The combination of frequent burning, draining, and damage due to other operations such as military manoeuvres and mineral extraction, have led to an irretrievable loss of this community in many areas and its distribution has been considerably fragmented with remaining stands closely hemmed in by coniferous plantations or intensive agricultural land.

M16

Molinia caerulea usually dominant with Scirpus cespitosus and Narthecium ossifragum less common than usual, scattered plants of Potentilla erecta and Succisa pratensis frequent, with occasional Polygala serpyllifolia, Carex panicea, Salix repens, Cirsium dissectum and Serratula tinctoria. Bryophytes usually sparse.

Scirpus cespitosus and Narthecium ossifragum generally frequent but listed associates of M16b usually uncommon.

Sphagnum carpet often very patchy and sometimes absent and M16c associates hardly ever found.

Very variable mixtures of Calluna vulgaris, Erica tetralix and Molinia caerulea generally dominate with occasional Scirpus cespitosus and Juncus squarrosus. Other associates listed for M16d rare.

M16a

Typical sub-community

All variations in the proportions of Molinia caerulea, Erica tetralix and Calluna vulgaris can be found in this sub-community so the appearance is very diverse. Where their cover is open, Sphagnum compactum and S. tenellum can be very frequent and often abundant. Among other bryophytes Hypnum jutlandicum and Kurzia pauciflora are the commonest species, while Campylopus brevipilus is very consistent in some areas. Lichens, particularly Cladonia impexa, can also be frequent.

This sub-community is found throughout the south of Britain and as more impoverished stands further north.

Molinia caerulea reduced in frequency and abundance and Calluna vulgaris often exceeding Erica tetralix. Frequent tussocks of Scirpus cespitosus and Juncus squarrosus. Hypnum cupressiforme, Dicranum scoparium, Racomitrium lanuginosum and Diplophyllum albicans all common. Cladonia species often present.

M16d

Juncus squarrosus – Dicranum scoparium sub-community

Molinia caerulea rarely has high cover and may be absent and although Erica tetralix retains constancy, Calluna vulgaris is usually dominant. Scirpus cespitosus is more frequent and has higher cover than in other sub-communities. Sphagnum compactum is often abundant and has a high frequency together with S. tenellum while S. capillifolium and S. subnitens are relatively scarce. Together with the bryophytes listed above lichens are often prominent, with Cladonia impexa and C. uncialis especially common.

This is the usual sub-community in the north and east of Britain.

Sub-shrub cover generally patchy, but Sphagnum carpet quite extensive with frequent Kurzia pauciflora. Wetter hollows and runnels have Drosera rotundifolia and often D. intermedia, Rhynchospora alba and locally R. fusca as sub-community preferentials.

M16c

Rhynchospora alba – Drosera intermedia sub-community

Molinia caerulea and Erica tetralix remain very frequent but Calluna vulgaris is less common and the cover of all three is reduced. In the intervening open areas is an extensive cover of Sphagnum compactum and S. tenellum with leafy hepatics and locally prominent Cladonia species, particularly C. impexa. There are scattered small tussocks of Scirpus cespitosus and often Narthecium ossifragum and Eriophorum angustifolium. More distinctive are the preferentials listed above.

This sub-community is concentrated in the New Forest and Poole Harbour.

M16b

Succisa pratensis – Carex panicea sub-community

Molinia caerulea tends to predominate with Erica tetralix and Calluna vulgaris having high frequencies. The vascular flora is richer than in M16a with Potentilla erecta and Succisa pratensis being constant and the species listed above preferential. Myrica gale sometimes shows local abundance. Bryophytes and lichens tend to be less common in this sub-community, but unusually Sphagnum auriculatum can be frequent and locally abundant.

This sub-community occurs throughout south-west England and perhaps elsewhere.

M17 *Scirpus cespitosus – Eriophorum vaginatum* blanket mire

This community is dominated by mixtures of monocotyledons, ericoid sub-shrubs and *Sphagnum* spp. It can occur as extensive, relatively uniform tracts, or as hummock and hollow complexes, with this community giving way to bog pool vegetation in the hollows. Among the bulkier vascular species, the most common are *Scirpus cespitosus*, *Eriophorum vaginatum*, *Molinia caerulea*, *Calluna vulgaris* and *Erica tetralix*; mixtures of which form a rather open uneven-topped tier which is 20-30 cm tall. *Myrica gale* also has occasional local abundance in this stratum.

Eriophorum angustifolium and *Narthecium ossifragum* are both very frequent and *Drosera rotundifolia* is very common in wetter areas. *Potentilla erecta* is a constant which helps to distinguish this community from other Sphagnetalia mires (M18-M21). Other species found at low frequencies throughout are *Pedicularis sylvatica*, *Huperzia selago*, *Juncus acutiflorus*, *Festuca ovina* and *Carex echinata*. *Vaccinium myrtillus*, *Empetrum nigrum* ssp. *nigrum* and *Rubus chamaemorus* are all scarce, in contrast to *Calluna vulgaris-Eriophorum vaginatum* blanket mire (M19).

Sphagnum spp. are an important component of the ground layer. *Sphagnum capillifolium* and *S. papillosum* are constants and may be accompanied by *S. tenellum*, *S. subnitens* and other species, forming luxuriant carpets. Such carpets typically have a variety of leafy hepatics including *Odontoschisma sphagni*, *Mylia anomala*, *M. taylori* and *Pleurozia purpurea*. *Racomitrium lanuginosum* is a common moss throughout, but becomes abundant on hummock tops and in degraded mires. Lichens, particularly larger *Cladonia* species, can be prominent and tend to be associated with *R. lanuginosum*.

This community is the characteristic blanket bog vegetation of the more oceanic parts of Britain, occurring extensively on waterlogged ombrogenous peat. The peats show varying humification but are typically highly acidic, with a surface pH usually not above 4 and often less. It is a community of lower altitudes where extreme humidity is combined with a relatively mild winter climate. It is largely confined to western Britain from the western Highlands of Scotland and the Western Isles, to south-west Scotland, the Lake District, Wales and south-west England.

Burning, marginal peat-cutting, and drainage have often resulted in surface drying of the peat and hence a modification of the vegetation. It is also possible that natural climatic change too has played a part in the degeneration of the blanket bogs occupied by this type of community. However, this community still remains as climax vegetation in the more oceanic parts of Britain.

M17

Erica tetralix and Molinia caerulea usually well represented with Drosera rotundifolia very frequent and Myrica gale locally plentiful. Extensive and varied Sphagnum carpet with leafy hepatics often prominent.

The usual vascular dominants are Calluna vulgaris and Scirpus cespitosus, with Erica tetralix and Molinia caerulea of reduced importance and Myrica gale rare. Drosera rotundifolia at most occasional. Sphagnum cover rather impoverished and leafy hepatics infrequent.

Racomitrium lanuginosum very common with several Cladonia species. Erica cinerea occasional and locally abundant. Vaccinium myrtillus and Empetrum nigrum very scarce. Nardus stricta and Juncus squarrosus occasional.

Juncus squarrosus, Nardus stricta and Deschampsia flexuosa frequent with small amounts of Vaccinium myrtillus and occasional Empetrum nigrum, but Erica cinerea rare. Racomitrium lanuginosum occasional and Cladonia species uncommon.

M17a

Drosera rotundifolia – Sphagnum spp. sub-community

This has a consistent representation of all the community constants. Among vascular dominants mixtures of Calluna vulgaris and Scirpus cespitosus or Calluna vulgaris and Molinia caerulea usually make up the bulk of the cover, with Eriophorum vaginatum sometimes showing local abundance in higher areas and Erica tetralix in wetter parts. Drosera rotundifolia is strongly preferential. Sphagnum spp. are extensive; Sphagnum capillifolium and S. papillosum are most abundant with S. compactum occasional and S. tenellum and S. subnitens frequent. Common leafy hepatics are Pleurozia purpurea and Odontoschisma sphagni. Racomitrium lanuginosum is also frequent but only as scattered shoots.

This sub-community occurs throughout the range of M17, but it is particularly extensive in north-west Scotland.

M17b

Cladonia spp. sub-community

Calluna vulgaris and Scirpus cespitosus are fairly consistent co-dominants with Molinia caerulea and Erica tetralix playing a subordinate role and Eriophorum vaginatum distinctly patchy. Myrica gale is scarce and Erica cinerea quite frequent and locally prominent. The Sphagnum carpet is much impoverished with Sphagnum capillifolium as the main species, often rather patchy, and all other species reduced in frequency. The leafy hepatics of M17a are uncommon, with Mylia taylori and Diplophyllum albicans preferential at low frequencies. Racomitrium lanuginosum and Cladonia spp. are increased in frequency and abundance, particularly Cladonia impexa, C. uncialis and C. arbuscula.

This sub-community along with the Juncus – Rhytidiadelphus sub-community occurs in the west but they extend the range of the community on to drier peats, most notably in south-west and eastern Scotland.

M17c

Juncus squarrosus – Rhytidiadelphus loreus sub-community

Calluna vulgaris and Scirpus cespitosus are the main vascular dominants with Erica tetralix and especially Molinia caerulea being reduced. Myrica gale is absent and Erica cinerea very scarce. There is a marked increase in Juncus squarrosus, Nardus stricta, Deschampsia flexuosa and Carex nigra, and with them Agrostis canina ssp. canina, Anthoxanthum odoratum and Luzula multiflora can be found. Of the Sphagnum spp., Sphagnum papillosum is usually the most abundant. There is a distinctive contingent of mosses: Hypnum cupressiforme/jutlandicum, Rhytidiadelphus loreus and Dicranum scoparium are all very frequent, while Polytrichum commune, P. alpestre, Plagiothecium undulatum, Aulacomnium palustre, Ptilidium ciliare, Pohlia nutans and Campylopus paradoxus are more occasional but preferential.

This sub-community occurs along with the Cladonia sub-community in the west but they extend the range of the community on to drier peats, most notably in south-west and eastern Scotland.

M18 *Erica tetralix – Sphagnum papillosum* raised and blanket mire

This community is generally dominated by *Sphagnum* spp. Ericoid sub-shrubs and monocotyledons are often subordinate. It can be found as undulating carpets or can comprise lawn and hummock components. The bulkier vascular plants typically form a low, patchy canopy with *Calluna vulgaris*, *Erica tetralix* and *Eriophorum vaginatum* being the commonest species and *Scirpus cespitosus* slightly less frequent. *Erica tetralix* tends to predominate on wetter ground where shoots of *Eriophorum angustifolium* can also be abundant. *Calluna*, *Scirpus* and *E. vaginatum* are found more typically on the drier areas.

Sphagnum spp. make up the most important component of the vegetation. Both *Sphagnum papillosum* and *S. capillifolium* are very common and *S. tenellum* is also a constant but less abundant. *Sphagnum magellanicum* is a preferential species and a major peat-builder. *S. imbricatum* is a distinctive species where present. Over gently-undulating surfaces the *Sphagnum* spp. form an irregular patchwork, but with increasing differentiation of hummocks and hollows they show a vertical stratification. On hummock tops and sides *S. capillifolium* is abundant, and *S. papillosum*, *S. magellanicum*, and a little *S. tenellum* predominate on the surrounds to wetter depressions.

Other bryophytes are generally subordinate but can be frequent and locally abundant. The leafy hepatics *Odontoschisma sphagni* and *Mylia anomala* are both common, but *Pleurozia purpurea* is generally absent. *Aulacomnium palustre* and *Hypnum cupressiforme/jutlandicum* are frequent mosses.

This vegetation is characteristic of waterlogged ombrogenous peats, typically at low altitudes in moderately oceanic parts of Britain. It is characteristic of raised bogs where it is the main community of the active plane, but is also found within blanket mires and in some basin mires on acid peat. The peats it covers are usually deep with a uniformly acid surface with a pH of about 4, and oligotrophic. It is widespread but local through the lowlands of Wales, up to the Scottish Borders and in south-west Scotland. There are also localities in southern England and east Scotland.

Erica tetralix – Sphagnum papillosum raised and blanket mire vegetation is a climax of a hydroseral succession. However, the typical habitat of this community has been widely affected by various treatments, notably peat-cutting, burning and grazing, and these have often modified the vegetation or reduced it to fragmentary stands.

M18

Sphagnum spp. luxuriant, with *Sphagnum magellanicum* frequent and abundant, along with *S. papillosum* in wetter lawns. *Narthecium ossifragum* and *Drosera rotundifolia* common and *Vaccinium oxycoccos* and *Andromeda polifolia* especially distinctive.

Sphagnum spp. abundant, with *Sphagnum capillifolium* usually dominant, *S. papillosum* frequent but usually subordinate and *S. magellanicum* only occasional. *Narthecium ossifragum* and *Vaccinium oxycoccos* occasional and *Drosera rotundifolia* and *Andromeda polifolia* very scarce. Sub-shrubs often quite vigorous. *Cladonia* species frequent and locally abundant.

M18a

Sphagnum magellanicum – Andromeda polifolia sub-community

All the community vascular constants are of high frequency but very often none is dominant. *Sphagnum* spp. form an obvious and extensive carpet in which *Sphagnum papillosum*, often with abundant *S. magellanicum*, predominates along with the constants *S. tenellum* and *S. capillifolium*. Scattered through are frequent individuals of *Drosera rotundifolia* and *Narthecium ossifragum* with the preferentials *Vaccinium oxycoccos* and *Andromeda polifolia*. *Cladonia* spp. and *Pleurozium schreberi* are typically of low frequency.

This sub-community occurs throughout the range of M18.

M18b

Empetrum nigrum ssp. nigrum – Cladonia spp. sub-community

Calluna vulgaris, *Scirpus cespitosus* and *Eriophorum vaginatum* tend to have higher covers here, *Calluna vulgaris* in particular becoming more vigorous and abundant. *Empetrum nigrum* is also frequent among them. Among the *Sphagnum* spp., *S. capillifolium* is strongly dominant, although *S. papillosum* is still frequent. *S. tenellum* is patchy and *S. magellanicum* only occasional. Other mosses become frequent, with *Pleurozium schreberi* and *Rhytidiadelphus loreus* being good preferentials; a range of hepatics is also common. There is a marked increase in *Cladonia* spp., notably *Cladonia impexa*, *C. uncialis* and *C. arbuscula*, each of which can be locally abundant.

This sub-community occurs throughout the range of M18.

43

M19 *Calluna vulgaris – Eriophorum vaginatum* blanket mire

This vegetation is generally dominated by mixtures of *Eriophorum vaginatum* and ericoid sub-shrubs. *Sphagnum* spp. can be prominent over wetter ground but are not as luxuriant or rich as in *Scirpus cespitosus – Eriophorum vaginatum* blanket mire (M17) or *Erica tetralix – Sphagnum papillosum* raised and blanket mire (M18). The ground surface is often uneven, but does not show true hummock and hollow relief. *Eriophorum vaginatum* is abundant and at least co-dominant. Normally this community has very frequent occurrences of *Calluna vulgaris*, *Vaccinium myrtillus* and *Empetrum nigrum* ssp. *nigrum* and, at higher altitudes, *V. vitis-idaea*, *V. uliginosum* and *E. nigrum* ssp. *hermaphroditum*. Overall *Calluna* is the most common co-dominant along with *Eriophorum vaginatum*, but diverse mixtures of these sub-shrubs are very frequent. Vascular associates are few, the commonest being *Eriophorum angustifolium*, and *Rubus chamaemorus*, a species which is distinctive for this community. *Deschampsia flexuosa* and *Juncus squarrosus* occur occasionally throughout, and at higher altitudes *Carex bigelowii* becomes frequent.

In contrast to the vascular plants, the bryophyte flora is rich, often with a cover exceeding 50%. The most frequent *Sphagnum* species is *S. capillifolium*, which forms patches rather than carpets. Hypnaceous mosses are consistently present and *Pleurozium schreberi*, *Rhytidiadelphus loreus*, *Hypnum cupressiforme/jutlandicum* and *Plagiothecium undulatum* are all very frequent. *Hylocomium splendens* is common at higher altitudes.

A variety of leafy hepatics occur in this community and lichens are frequent. Larger *Cladonia* spp. can be abundant on old *Eriophorum* hummocks.

This mire is the typical blanket bog vegetation of high-altitude ombrogenous peats present in the wet and cold climate of the uplands of northern Britain. In particular, it occurs on high-level plateaux and broad watersheds, usually above 300 m, and is confined to deeper peats, usually more than 2 m thick, on flat or gently-sloping ground. The peats are usually well-humified, highly acidic with a surface pH often less than 4. They are not consistently waterlogged and may become surface oxidised in summer. Erosion of the peat is common. This community is found on the higher ground in the Pennines, the central Highlands of Scotland, and Wales.

Treatments such as burning and grazing are important in influencing the composition and structure of the vegetation throughout the range of this community, in particular where stands form part of unenclosed hill grazing or grouse moors. A stable diversity of bog vegetation can be maintained by careful burning on a rotation of around 10 years, or by moderate levels of grazing. However, frequent burning or heavy grazing contribute to the conversion of the *Calluna vulgaris – Eriophorum vaginatum* blanket mire to *Eriophorum vaginatum* blanket mire (M20). In other cases drainage can convert this community into heathland or grassland, and this type of blanket mire has been reclaimed for agriculture or forestry in many areas.

M19

Eriophorum vaginatum usually abundant and *Rubus chamaemorus* frequent. *Scirpus cespitosus* rare and *Sphagnum* cover patchy.

Empetrum ssp. *hermaphroditum* and *Vaccinium vitis-idaea* constant and often co-dominant with *Calluna vulgaris*, *V. myrtillus* and *Eriophorum vaginatum*. *Hylocomium splendens*, *Racomitrium lanuginosum* and *Polytrichum alpestre* common in the ground layer.

M19c

Vaccinium vitis-idaea – Hylocomium splendens sub-community

This embraces all the high-montane blanket mire. It preserves all the general floristic features but is distinctive in the frequent and consistent presence of *Vaccinium vitis-idaea* and *Empetrum nigrum* ssp. *hermaphroditum* and the more restricted presence of *V. uliginosum*. These are mixed with other sub-shrubs and usually with abundant *Calluna vulgaris*. *Eriophorum angustifolium* is less common, but *Rubus chamaemorus* is very frequent. *Juncus squarrosus* occurs quite often, as do *Carex bigelowii* and *Scirpus cespitosus*. *Sphagnum* spp. can be quite prominent, with a range of other bryophytes including the hepatic *Ptilidium ciliare*. The lichen flora is usually well developed.

This sub-community extends to the altitudinal limit of this kind of blanket bog in central Scotland.

Three variants.

Empetrum nigrum ssp. *hermaphroditum* occasional at most and often with ssp. *nigrum* as a replacement. *Vaccinium vitis-idaea* also occasional and *V. uliginosum*, *Juncus squarrosus* and *Carex bigelowii* rare. Bryophytes in M19c rare.

M19b

Empetrum nigrum ssp. *nigrum* sub-community

This sub-community preserves all the features of M19 although it is very variable in all its structural elements. *Eriophorum vaginatum* is usually abundant, with sub-shrubs playing a consistent role and *Calluna vulgaris* being the leading species, though *Vaccinium myrtillus* and/or *Empetrum nigrum* ssp. *nigrum*, particularly the latter, can increase greatly after burning. *Erica tetralix* is very scarce and *Scirpus cespitosus* and *Molinia caerulea* have a low frequency. *Rubus chamaemorus* however increases greatly. The *Sphagnum* flora is impoverished and on drier ground hypnaceous mosses are particularly abundant.

This forms the richer type of 'Pennine blanket bog' extending northwards through Cheviot and the Borders into eastern Scotland.

Eriophorum vaginatum less abundant, *Erica tetralix* more common and *Scirpus cespitosus* frequent. *Rubus chamaemorus* rare. *Sphagnum* spp. cover quite extensive.

M19a

Erica tetralix sub-community

Eriophorum vaginatum can be co-dominant with the sub-shrubs though generally less abundant than in other sub-communities. *Calluna vulgaris* is often the predominant sub-shrub with some *Empetrum nigrum* ssp. *nigrum*. *Erica tetralix* is preferential and sometimes with high cover. *Scirpus cespitosus* is frequent and locally abundant. The infrequency of *Rubus chamaemorus* is distinctive. *Sphagnum* spp. tend to be consistently abundant with *S. capillifolium* being commonly accompanied by *S. papillosum*.

This is the usual form found at lower altitudes that have a rather more oceanic climate, such as in Wales and south-west Scotland.

M20 *Eriophorum vaginatum* blanket and raised mire

Eriophorum vaginatum mire comprises species-poor ombrogenous bog vegetation dominated by *E. vaginatum*, the tussocks of which form an open or closed canopy 10-30 cm high. The only other constant plant is *E. angustifolium*, which is usually found as scattered shoots. Ericoid sub-shrubs are patchy; *Calluna vulgaris*, *Empetrum nigrum* ssp. *nigrum* and *Vaccinium myrtillus* can each be found quite frequently and the last two may be locally abundant. Alternatively these species may be reduced to sparse shoots. *Deschampsia flexuosa* is fairly common while *Festuca ovina*, *Juncus squarrosus*, *Scirpus cespitosus* and *Carex bigelowii* are all infrequent.

Bryophytes are sparse and patchy. *Sphagnum* spp. are scarce with *Sphagnum capillifolium* and *S. papillosum* the most usual species. Hypnaceous mosses are poorly represented; the only moss of any frequency being *Campylopus paradoxus* which can be accompanied by *Dicranum scoparium*. There may be occasional shoots of *Orthodontium lineare*, *Pohlia nutans* and *Drepanocladus fluitans*. A variety of leafy hepatics may be present. Lichens are typically few in number.

Bulkier species like *Cladonia arbuscula*, *C. uncialis* and *C. impexa* can sometimes be found or sometimes there is just a patchy cover of peat encrusters.

This community is characteristic of ombrogenous peats on bogs where certain treatments have greatly affected the vegetation; grazing and burning have been of greatest significance, but draining and aerial pollution have also played a part. It is commonest on blanket mires, where these factors have contributed both to floristic impoverishment and to gross erosion of the peats, but is also found locally on run-down raised bogs. The *Eriophorum* mire is present mainly between 500 m and 700 m where the climate is cold and wet. The peats are generally dry, often showing surface oxidation and with a pH frequently as low as 3. This community can be found locally through northern Britain, and is especially extensive in the southern Pennines.

This community has been seen to revert to the vegetation characteristic of the richer blanket bog community *Calluna vulgaris* – *Eriophorum vaginatum* mire (M19) within 25 years of enclosure and freedom from burning, but in many instances, particularly if intensive grazing or frequent burning have been accompanied by drainage, the degeneration of the vegetation is perhaps irreversible.

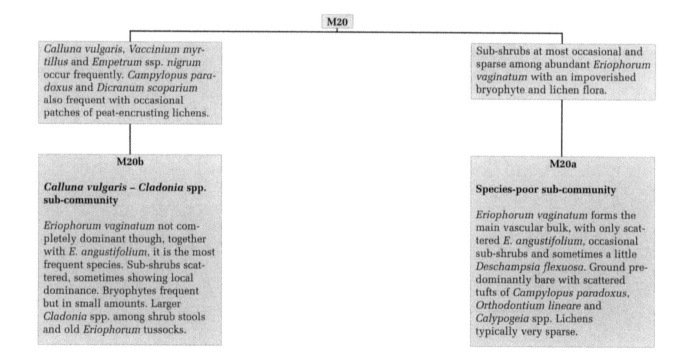

M20

Calluna vulgaris, *Vaccinium myrtillus* and *Empetrum* ssp. *nigrum* occur frequently. *Campylopus paradoxus* and *Dicranum scoparium* also frequent with occasional patches of peat-encrusting lichens.

Sub-shrubs at most occasional and sparse among abundant *Eriophorum vaginatum* with an impoverished bryophyte and lichen flora.

M20b

Calluna vulgaris – Cladonia spp. sub-community

Eriophorum vaginatum not completely dominant though, together with *E. angustifolium*, it is the most frequent species. Sub-shrubs scattered, sometimes showing local dominance. Bryophytes frequent but in small amounts. Larger *Cladonia* spp. among shrub stools and old *Eriophorum* tussocks.

M20a

Species-poor sub-community

Eriophorum vaginatum forms the main vascular bulk, with only scattered *E. angustifolium*, occasional sub-shrubs and sometimes a little *Deschampsia flexuosa*. Ground predominantly bare with scattered tufts of *Campylopus paradoxus*, *Orthodontium lineare* and *Calypogeia* spp. Lichens typically very sparse.

M21 *Narthecium ossifragum – Sphagnum papillosum* valley mire

This mire vegetation is dominated by carpets of *Sphagnum* spp. with scattered herbs and sub-shrubs forming extensive lawns or the drier parts of hummock and hollow complexes. The dominant *Sphagnum* is usually *S. papillosum*. *Sphagnum auriculatum* or *S. recurvum* (and occasionally both) are frequently encountered and less often *S. cuspidatum*. Locally, *S. magellanicum* and *S. pulchrum* may be present. *Sphagnum compactum* is almost always absent in contrast with *Erica tetralix – Sphagnum compactum* wet heath (M16). Only a few other mosses occur and they are generally of low cover, but leafy hepatics are common. *Odontoschisma sphagni* and *Kurzia pauciflora* are most common, but *Cladopodiella fluitans*, *Cephalozia macrostachya*, *C. connivens*, *C. bicuspidata* and *Calypogeia fissa* have also been recorded. Lichens are typically sparse, but hummock tops can be a habitat for *Cladonia impexa*, *C. arbuscula* and *C. uncialis*.

The vascular plant species present provide the major distinction between this community and other types of Erico – Sphagnion bogs (M17-M20). *Eriophorum vaginatum* and *Scirpus cespitosus* are rare, but *Eriophorum angustifolium* and *Narthecium ossifragum* are constant and often abundant. *Molinia caerulea* is typical, but is often weak and not tussock-forming, and *Rhynchospora alba* is characteristic of one sub-community, being most prominent around hollows and pools. The sub-shrubs *Erica tetralix* and *Calluna vulgaris* are very frequent, forming an open canopy. More restricted but conspicuous when present is *Myrica gale*. Other plants are generally present as scattered individuals, *Drosera rotundifolia* being the most frequent.

This is a community of permanently waterlogged, acid and oligotrophic peats, especially characteristic of valley mires maintained by a locally high water table. The peat on which this community is found is usually not very deep (20-150 cm) with a surface pH mostly in the range of 3.5-4.5, and a water table at or very close to the surface. It is found locally in the southern lowlands of Britain.

Neither burning nor grazing are important in maintaining this community, as the wetness of the vegetation gives its some protection from these treatments, but both can cause damage if combined with drainage. With continued autogenic accumulation of peat, it would probably progress naturally to some kind of woodland but this process is likely to be very slow in the absence of drainage.

```
                              M21
    ┌──────────────────────────┴──────────────────────────┐
```

Rhynchospora alba very frequent and *Myrica gale* occasional, sometimes abundant in the *Sphagnum* carpet, which is usually dominated by mixtures of *S. papillosum* and *S. auriculatum*, with *S. recurvum* scarce.	*Rhynchospora alba* and *Myrica gale* rather infrequent, but *Vaccinium oxycoccos* patchily present and *Potentilla erecta* occasional in a carpet usually dominated by *Sphagnum recurvum* with some *S. papillosum* but little *S. auriculatum*.

M21a	**M21b**
Rynchospora alba – Sphagnum auriculatum sub-community	**Vaccinium oxycoccos – Sphagnum recurvum sub-community**
In this, the most frequently described M21 community, the *Sphagnum* carpet is generally dominated by mixtures of *S. papillosum* and *S. auriculatum*. Hepatics are varied and often abundant. All vascular constants have a high frequency and in addition *Rhynchospora alba* is very frequent. *Myrica gale* frequently shows local abundance.	*Sphagnum papillosum* is often abundant but somewhat patchy and *S. recurvum* frequently has an equal cover. *S. auriculatum* is much reduced in frequency. *Odontoschisma sphagni* occurs sometimes but hepatics are greatly reduced. Among vascular plants *Rhynchospora alba* is scarce and *Vaccinium oxycoccos* reduced in frequency.
This sub-community is best represented in central southern England.	This sub-community extends into the north and west of England and Wales.

M22 *Juncus subnodulosus – Cirsium palustre* fen-meadow

This community shows considerable variation in its floristic composition which often reflects a unique and complex history of mowing and grazing management. The most prominent structural element typically consists of rushes and sedges of moderate stature, appearing as a rank sward if it is not grazed. *Juncus subnodulosus* is the most important of the bulkier species and the most usual dominant. *Juncus inflexus* and *J. articulatus* are the commonest accompanying rushes. Of the sedges, the most striking are *Carex acutiformis* and *C. disticha*, either of which can be frequent or occasionally dominant. Much more occasionally *C. elata* or *C. paniculata* can occur. In summer this rush and sedge layer can be overtopped by flowering dicotyledons. The most frequent of these throughout the community are *Cirsium palustre*, *Filipendula ulmaria*, *Angelica sylvestris*, *Succisa pratensis*, *Eupatorium cannabinum* and *Scrophularia aquatica*. More locally *Lythrum salicaria*, *Lysimachia vulgaris*, *Valeriana officinalis*, *Thalictrum flavum* and *Symphytum officinale* can be found, and, in Broadland, the nationally rare *Peucedanum palustre*. However, grazing may keep these species severely in check.

Among smaller species, grasses are important, and species found include *Holcus lanatus*, *Festuca rubra* and, less commonly, *Arrhenatherum elatius*, together with *Poa trivialis*, *Agrostis stolonifera*, *Anthoxanthum odoratum* and *Briza media*. Smaller herbs are those characteristic of moist grasslands, the commonest being *Mentha aquatica*, *Caltha palustris*, *Equisetum palustre*, *Carex panicea*, *Valeriana dioica*, *Hypericum tetrapterum*, *Hydrocotyle vulgaris*, *Lotus uliginosus*, *Lychnis flos-cuculi*, *Cardamine pratensis*, *Ranunculus acris*, *Potentilla erecta*, *Cerastium fontanum*, and *Rumex acetosa*, with several scrambling plants including *Galium uliginosum*, *G. palustre*, *Vicia cracca* and *Lathyrus pratensis*. Bryophytes play a reduced role with only *Calliergon cuspidatum* and *Brachythecium rutabulum* being common throughout.

This community brings together secondary herbaceous vegetation developed over a variety of moist, base-rich and moderately mesotrophic peats in southern lowland Britain. It can be found either in, or around, well-developed springs, flushes and mires, or marking out more ill-defined areas of influence of surface or ground waters. This community marks out soils which are kept moist for most of the year and have a moderate to high base-status, and usually a pH range of 6.5-7.5. The community has a wide distribution through the southern British lowlands with particular concentrations of stands in East Anglia, north Buckinghamshire and Anglesey.

This community is always dependent on various kinds of treatment for its maintenance, as it is derived from other wetland vegetation types by mowing or grazing, or both. The reduction in grazing results in the expansion of bulky dominants and ranker grasses and the overwhelming of the smaller herbs. Its overall distribution and the extent of the stands has become much less than formerly because of intensive land improvement and abandonment of traditional agricultural practices.

M22

Phragmites australis, Lythrum salicaria and Hydrocotyle vulgaris frequent, with Carex acutiformis or C. elata common, sometimes abundant.

Species listed opposite at most occasional and hardly ever abundant.

Juncus subnodulosus usually dominant in rather rank and impoverished vegetation with bulkier grasses abundant. Molinia caerulea and dicotyledons listed opposite infrequent at most.

Juncus subnodulosus often accompanied by J. inflexus and J. articulatus; also Carex disticha and Deschampsia cespitosa patchily abundant. Molinia caerulea, Briza media and Anthoxanthum odoratum common at low cover with frequent records for Cardamine pratensis, Cerastium fontanum, Trifolium repens, T. pratense, Rumex acetosa, Plantago lanceolata and Epilobium parviflorum.

Carex elata constant and sometimes abundant among Juncus subnodulosus, with frequent Potentilla palustris, Epilobium palustre, Equisetum fluviatile and Dactylorhiza incarnata.

Juncus subnodulosus usually dominant with little or no Carex elata, but with frequent and sometimes abundant Molinia caerulea and C. acutiformis. Galium palustre common but other associates listed opposite rare.

M22a

Typical sub-community

In this, the most common kind of fen-meadow, there are no preferential floristic features and the general impression is of rank structure and impoverishment. Juncus subnodulosus is frequent and the commonest dominant. The commonest associates are bulkier grasses such as Holcus lanatus, Festuca rubra, sometimes Molinia caerulea, mat-formers such as Agrostis stolonifera and Poa trivialis, and taller dicotyledons such as Cirsium palustre, Filipendula ulmaria, Angelica sylvestris, Succisa pratensis and Eupatorium cannabinum. There are also sprawling plants and a few shorter species such as Mentha aquatica and Equisetum palustre. Within this general framework there is considerable local variation.

This sub-community is the most frequent type of spring-fed stands.

M22b

Briza media – Trifolium spp. sub-community

In comparison to M22a, in this sub-community the rush and sedge tier has a lower cover and density and there is a richer associated flora. Taller dicotyledons such as Cirsium palustre, Filipendula ulmaria, Angelica sylvestris, Eupatorium cannabinum and Succisa pratensis show an increased frequency and often occur with Centaurea nigra and Rumex acetosa. The lower sward is particularly rich. There is a range of grasses, often with Carex panicea and occasionally C. nigra and C. flacca. There is a variety of other herbs: most frequent are Lotus uliginosus, Lychnis flos-cuculi, Caltha palustris, Ranunculus acris, Valeriana dioica, Potentilla erecta. P. anserina and Hypericum tetrapterum which are characteristic of M22 as a whole. Strongly preferential are the species listed above together with Prunella vulgaris, Ranunculus repens, Triglochin palustre, Rhinanthus minor and Dactylorhiza fuchsii. Phragmites australis is generally scarce.

This sub-community is most frequently developed around grazed spring-fens and wet field hollows and occurs through the range of M22.

M22c

Carex elata sub-community

Carex elata is uncommon in fen-meadow vegetation but here it occurs with some abundance and occasionally is co-dominant with Juncus subnodulosus. Phragmites australis is often present and there is usually some Hydrocotyle vulgaris. The plants listed above are preferential. Some stands have Pedicularis palustris, Menyanthes trifoliata, Ranunculus flammula, R. lingua and Berula erecta. Grasses and smaller herbs of M22b are particularly sparse.

This and the Iris sub-community are especially well-represented on topogenous mires in East Anglia.

M22d

Iris pseudacorus sub-community

Juncus subnodulosus is dominant, but as in M22c, Phragmites australis and Hydrocotyle vulgaris are often present and Menyanthes trifoliata, Potentilla palustris and Equisetum fluviatile occasional. Grasses and smaller herbs of M22b are sparse. Carex elata is occasional and tussocks of C. paniculata are sometimes prominent. Tall dicotyledons are prominent in summer: Cirsium palustre, Filipendula ulmaria, Angelica sylvestris and Succisa pratensis are all common and Iris pseudacorus, Ranunculus flammula, Valeriana officinalis, Lysimachia vulgaris and Thalictrum flavum are all frequent.

This and the Carex sub-community are especially well-represented on topogenous mires in East Anglia.

M23 *Juncus effusus/acutiflorus – Galium palustre* rush-pasture

This vegetation is ill-defined and characterised by the abundance of either *Juncus effusus* or *J. acutiflorus*, sometimes both, in a ground of mesophytic herbs common in moist agricultural grassland. The rushes often have a high cover but they may also be more sparse. *Juncus effusus* is more abundant in the east, while *J. acutiflorus* has a distinctly western distribution. Diversity in dominants is not great but the associates are quite diverse, making the bounds of this vegetation type hard to fix. Sometimes *J. articulatus* may be locally frequent and abundant. *Holcus lanatus* is the most frequent grass, but *Agrostis canina* ssp. *canina*, *A. stolonifera*, *Anthoxanthum odoratum* and *Poa trivialis* are also common in drier stands. *Festuca rubra* and *Agrostis capillaris* may also become frequent. *Molinia caerulea* is increasingly common and abundant towards the west and there may be some sedges in the sward. There is a variety of common herbs. Among the taller species *Cirsium palustre* is the commonest, *Rumex acetosa*, *Angelica sylvestris* and *Epilobium palustre* are frequent, and two sprawling species *Galium palustre* and *Lotus uliginosus* are constant. Frequent smaller species are *Mentha aquatica*, *Ranunculus flammula*, *R. repens*, *R. acris*, *Cardamine pratensis*, *Hydrocotyle vulgaris*, *Viola palustris*, and *Stellaria alsine*.

Bryophytes are variable in their cover. Where the vegetation is open they may be abundant. *Calliergon cuspidatum* is the most frequent, often occurring with *Brachythecium rutabulum* and *Rhytidiadelphus squarrosus*.

This rush-pasture occurs over a variety of moist, moderately acid to neutral, peaty and mineral soils in the cool and rainy lowlands of western Britain. It is a community of gently-sloping ground around the margins of soligenous flushes, as a zone around topogenous mires and wet heaths, and especially widespread in ill-drained, comparatively unimproved or reverted pasture. It can be found on a variety of moderately acid to neutral soils that are kept moist to wet for most of the year with a pH in the range of 4-6. It is found at the opposite climatic and edaphic extreme to the *Juncus subnodulosus – Cirsium palustre* fen meadow (M22) with a distinctly oceanic distribution. The community is wide-spread through the west of Britain from Devon and Cornwall to Skye and Caithness.

This community is maintained mainly by grazing and more occasionally mowing which prevents the succession of the community to woodland. Draining and other kinds of soil improvements such as fertilising and reseeding have reduced its former extent.

M23

Juncus effusus very common, but exceeded by *J. acutiflorus*, with *Molinia caerulea* and *Holcus lanatus* frequent and sometimes abundant. *Filipendula ulmaria* occasional, also some of *Ranunculus acris*, *Potentilla erecta*, *Achillea ptarmica* and *Equisetum palustre* and locally prominent tall-fen herbs such as *Lythrum salicaria* and *Iris pseudacorus*.

M23a

Juncus acutiflorus sub-community

Juncus acutiflorus and *J. effusus* are both constant, *J. articulatus* is locally prominent and *J. conglomeratus* is occasional. The commonest grasses are *Holcus lanatus* and, preferential here, *Molinia caerulea*. Community herbs such as *Cirsium palustre*, *Rumex acetosa* and *Angelica sylvestris* remain frequent with *Galium palustre* and *Lotus uliginosus*. *Filipendula ulmaria* is more common than in M23b and there may be an abundance of taller herbs. In the lower tier of vegetation, *Mentha aquatica*, *Cardamine pratensis*, *Ranunculus flammula* and *R. repens* are frequent with at least some of the herbs listed above.

This, the more sharply-defined sub-community, prevails in Scotland and is common in Wales.

Juncus effusus constant and usually dominant, with *J. acutiflorus* scarce. *Holcus lanatus* common, but *Molinia caerulea* and dicotyledons listed opposite all scarce.

M23b

Juncus effusus sub-community

This sub-community is less well-defined and is essentially a transition between M23a and the *Holcus lanatus – Juncus effusus* rush-pasture (MG10). Other rushes comprise *Juncus articulatus* and *J. conglomeratus*, but *J. inflexus*, common in MG10, is absent. *Molinia caerulea* is infrequent, but grasses are important in the sward. Where the community is surrounded by improved pasture, as is common, *Cynosurus cristatus* and *Lolium perenne* can be present at low cover. Good distinguishing features are the high frequencies of *Galium palustre*, *Cirsium palustre*, *Ranunculus flammula* and *Mentha aquatica*.

This is the typical sub-community of the South-West Peninsula, but is also found through the range of M23.

M24 *Molinia caerulea – Cirsium dissectum* fen-meadow

This community includes the bulk of the *Molinia caerulea* vegetation in the lowland south-east of Britain. *Molinia* is always the dominant to the extent that associates may be reduced to scattered individuals of only a few species. Often however there are a considerable number of associates. In structural terms the most important species are other monocotyledons. Through most of the central and eastern part of the range where this community is often found in association with fens, *Juncus subnodulosus* is the characteristic rush with *J. articulatus* and *J. inflexus* sometimes also present. To the south and west where the community often develops among heath vegetation, *J. acutiflorus* and *J. conglomeratus* become frequent.

The associated flora of dicotyledons helps to distinguish this community, although it is often difficult to separate it from *Juncus subnodulosus – Cirsium palustre* fen meadow (M22) and *Juncus effusus/acutiflorus – Galium palustre* rush pasture (M23) when these contain *Molinia. Cirsium palustre* and *Angelica sylvestris* are both very frequent and *Filipendula ulmaria* and *Centaurea nigra* can also be common. More strictly limited are *Valeriana dioica*, *Succisa pratensis*, and *Cirsium dissectum*, although only the last species is preferential. Other common species of wide occurrence are *Potentilla erecta*, *Lotus uliginosus*, *Mentha aquatica*, *Prunella vulgaris*, *Ranunculus acris*, *Hydrocotyle vulgaris*, and the scramblers *Vicia cracca* and *Lathyrus pratensis*.

Coarser grasses are often prominent, with *Holcus lanatus* and *Anthoxanthum odoratum* most frequent and *Festuca rubra*, *Deschampsia cespitosa* and *Agrostis stolonifera* less common, though sometimes abundant. There can also be some sedges present, the most common being smaller species such as *Carex panicea*, which is a community constant, and less frequent *C. hostiana* and *C. pulicaris*. Bryophytes are generally few and of low cover.

This is a community of moist to fairly dry peats and peaty mineral soils which are circumneutral, generally having a pH within the range 5-6.5. It can be found in association with both soligenous and topogenous mires, typically marking out the better-drained fringes of bogs and fens, or the margins of wet hollows and flushes. This community is widespread through the lowland south of Britain but has become increasingly local with changes in agricultural practice.

Although climate and soil together both influence the floristics of this community it is essentially a secondary vegetation type, derived from a variety of wetland vegetation types and maintained by mowing or grazing. In the absence of any kind of treatment all the stands of the community would probably progress to scrub or woodland. It has been reduced in extent by agricultural reclamation. Other stands have become rank and scrubby with neglect.

M24

Molinia caerulea generally dominant with Juncus subnodulosus common and frequent records for some of Valeriana dioica, Galium uliginosum, Centaurea nigra, Vicia cracca, Filipendula ulmaria and Equisetum palustre.

Juncus subnodulosus absent, but J. acutiflorus and J. conglomeratus common. J. effusus occasional. Species listed opposite all scarce but Erica tetralix, Calluna vulgaris, Galium palustre and Dactylorhiza maculata frequent.

M24c

Juncus acutiflorus – Erica tetralix sub-community

Rushes are a common feature, most usually Juncus acutiflorus and J. conglomeratus. There is frequently some Erica tetralix, less often Calluna vulgaris and Ulex gallii, so that the vegetation looks more like a heath. However Holcus lanatus, Anthoxanthum odoratum, less often Festuca rubra and Agrostis stolonifera are intermixed with Molinia caerulea with frequent Carex panicea and a range of small herbs. Valeriana dioica, Centaurea nigra and Filipendula ulmaria and the tall-fen herbs of M24a are not represented. Bryophytes are not conspicuous, though a number of species have been recorded.

This is the most usual type of M24 in south-western Britain.

Phragmites australis constant, Cladium mariscus quite common, both typically subordinate in cover to Molinia caerulea. Eupatorium cannabinum and Lythrum salicaria frequent.

Eupatorium cannabinum and Phragmites australis only occasional. Cladium mariscus and Lythrum salicaria usually absent.

M24b

Typical sub-community

Molinia caerulea is often found with smaller amounts of Juncus subnodulosus or J. articulatus. Smaller grasses are well represented: Holcus lanatus and Anthoxanthum odoratum are frequent and Briza media is strongly preferential. Sedges are common with Carex panicea and C. hostiana showing a peak of occurrence, C. pulicaris especially frequent and C. nigra preferential and sometimes in abundance. Succisa pratensis, Cirsium dissectum, C. palustre and Angelica sylvestris are all very common together with the species listed above. Sub-shrubs are typically sparse and bryophytes are poorly represented.

This is the most common sub-community in central and eastern England.

M24a

Eupatorium cannabinum sub-community

One of the most conspicuous species here, even when sparse, is Phragmites australis. The best single preferential is Eupatorium cannabinum which provides continuity with tall-herb fen; less common are Lythrum salicaria and Lysimachia vulgaris. Cirsium palustre, Angelica sylvestris and Filipendula ulmaria are also common and, as reed or sedge cover becomes thinner, plants such as Succisa pratensis, Cirsium dissectum, Centaurea nigra and Equisetum palustre increase in frequency. Bryophytes can be more conspicuous in this sub-community, but species are only few, with Campylium stellatum joining Calliergon cuspidatum and Brachythecium rutabulum as a distinctive preferential.

This sub-community is mainly confined to East Anglia.

M25 *Molinia caerulea – Potentilla erecta* mire

Molinia caerulea is the most abundant species found in this community The associated flora is poor, and most common are rushes and a few dicotyledons. Among the former, *Juncus acutiflorus* and *J. effusus* are the most frequent. *Juncus articulatus* and *J. subnodulosus* are both occasional, and *J. conglomeratus* is very scarce. The only constant dicotyledon is *Potentilla erecta*. *Lotus uliginosus*, *Succisa pratensis*, *Cirsium palustre* and *Angelica sylvestris* are sparse and occasionally there can be some *Eupatorium cannabinum* or *Filipendula ulmaria*. *Cirsium dissectum* is very rare and its presence separates the *Molinia caerulea – Cirsium dissectum* mire (M24) from this community. Also, since the soil pH is generally acidic, plants such as *Carex hostiana*, *C. pulicaris* and *Briza media*, frequent in M24, are of very limited occurrence here. Occasionally sub-shrubs can be quite common, particularly *Calluna vulgaris* and *Erica tetralix*. *Ulex gallii* can also be occasional in Wales and south-west England, and *U. europaea* occurs in some stands. *Myrica gale* is local but can be quite extensive and dense.

Grasses are limited in importance but *Agrostis canina* and *A. stolonifera* can be found at low frequency throughout and *Holcus lanatus* is fairly common. Among the dense herbage, bryophytes are sparse.

This mire is a community of moist, but well-aerated, acid to neutral peats and peaty mineral soils in the wet and cool western lowlands of Britain. It occurs over gently-sloping ground, marking out seepage zones and flushed margins of sluggish streams, water-tracks and topogenous mires, but also extends onto the fringes of ombrogenous mires. Soil and drainage conditions of this community have similarities to those of M23 and M24 and geographically this community can be seen as a northern/western replacement of M24. It is especially frequent in south-west England, Wales, and southern Scotland.

Although both climate and soils influence the composition of the vegetation, treatments such as burning, grazing and drainage are likely to be largely responsible for the development of this community over ground that would naturally carry some other kind of mire or wet heath vegetation. Tracts of this community have been replaced by coniferous plantations, particularly in the upland fringes of the north-west. Elsewhere in the lowlands, other tracts of the community together with neighbouring vegetation have been lost to agricultural improvements.

M25

Erica tetralix constant, Calluna vulgaris and Myrica gale quite frequent, with Eriophorum angustifolium common and occasional Narthecium ossifragum, Drosera rotundifolia and Vaccinium oxycoccos.

Juncus acutiflorus and occasionally J. effusus patchily prominent in a grassy community with frequent Holcus lanatus, Festuca rubra, Anthoxanthum odoratum, Agrostis capillaris, Danthonia decumbens, Luzula multiflora and L. campestris. Erica tetralix rare but Calluna vulgaris is occasional.

Species listed opposite usually all occasional but tall herbs prominent among Molinia caerulea and rush clumps, with frequent Angelica sylvestris and Cirsium palustre. Epilobium palustre, Eupatorium cannabinum, Filipendula ulmaria, Galium palustre and Mentha aquatica are occasional. Schoenus nigricans can be locally abundant.

M25a

Erica tetralix sub-community

This is the widely distributed type of M25. Molinia caerulea is the dominant but the associated flora is shifted towards that of Erica tetralix wet heaths (M15-16). Erica tetralix is strongly preferential, with frequent Calluna vulgaris. Juncus acutiflorus remains common but is joined by Eriophorum angustifolium. Apart from sparse Anthoxanthum odoratum, Festuca rubra and Agrostis canina, grasses are thin and taller herbs are poorly represented. Among smaller plants, Viola palustris and Hydrocotyle vulgaris are sometimes present, but with occasional Narthecium ossifragum, Drosera rotundifolia and Vaccinium oxycoccos. Bryophytes are distinctive, with Aulacomnium palustre, Polytrichum commune, Hypnum jutlandicum and Calypogeia fissa all preferential. Sphagnum spp. are noticeable, forming patches; Sphagnum recurrum and S. auriculatum are the commonest species.

This sub-community can be found throughout the range of M25.

M25b

Anthoxanthum odoratum sub-community

Although Molinia caerulea is still dominant, the sward is shorter and more varied. Apart from the rushes and grasses mentioned above other associates can be sparse. Calluna vulgaris and Ulex galliiare occasional, as are U. europaeus and Rubus fruticosus. Erica tetralix and Myrica gale are very uncommon. Succisa pratensis, Lotus uliginosus and Cirsium palustre are all more frequent than in M25a and Serratula tinctoria and Rumex acetosa are weakly preferential, but all tend to be grazed to rosettes. Smaller herbs and bryophytes are poorly represented.

This sub-community is scattered throughout the range of M25 but is particularly frequent in Wales.

M25c

Angelica sylvestris sub-community

This is the most local sub-community, but also the most striking sub-community, developed on moist ground with freedom from grazing. Molinia caerulea is vigorous, but variegated by clumps of Juncus acutiflorus and J. effusus. Taller dicotyledons are common: Succisa pratensis and Lotus uliginosus are notable together with the preferential species listed above and also Pulicaria dysenterica, Valeriana officinalis and Centaurea nigra. Shorter species include Mentha aquatica, Cardamine pratensis and Equisetum palustre with frequent Galium palustre. Bryophytes are again sparse but Calliergon cuspidatum and C. giganteum form scattered patches.

This sub-community is found mainly in south-west England and south-west Wales.

M26 *Molinia caerulea – Crepis paludosa* mire

This community is well-defined by a block of constants and frequent companions but also shows considerable variation in associated flora. Stands range from swamp to those having a rank, grassy character. *Molinia caerulea* is almost always present, being the commonest dominant. *Carex nigra* is also a constant, often as prominent tufts which can exceed *Molinia* in cover. *Carex panicea* can be abundant and *C. pulicaris* is common. In stands transitional to swamp, *C. appropinquata* or *C. rostrata* are present. In the *Festuca* sub-community, by contrast, it is taller rushes and grasses which, with *Molinia*, form the bulk of the cover. *Juncus acutiflorus* may form dense patches and *J. conglomeratus* and *J. articulatus* both occur occasionally.

Hemicryptophyte dicotyledons are an important structural element among the *Molinia*, sedges and rushes. Most frequent are *Succisa pratensis*, *Filipendula ulmaria*, *Valeriana dioica*, *Cirsium palustre* and *Caltha palustris* together with the northern species *Crepis paludosa* and *Trollius europaeus*. Also common are *Sanguisorba officinalis*, *Angelica sylvestris*, *Centaurea nigra*, *Leontodon hispidus*, *Geum rivale* and *Lychnis floscuculi*. Less conspicuous, but also frequent, is *Equisetum palustre*. *Potentilla erecta*, *Ranunculus acris*, and *Anemone nemorosa* are common. Bryophytes are only prominent in exceptional cases, with *Calliergon cuspidatum* most frequent.

This is a very local community of moist, moderately base-rich and calcareous peats and peaty mineral soils in the sub-montane northern Pennines. It represents a northern and altitudinal extreme of the richer kind of *Molinia* – tall herb vegetation. Stands are rare but all occur around the northern Pennine uplands and the Lake District between 250 m and 450 m altitude.

This community is an apparently stable component of topogenous sequences around open waters and mires, but where it occurs on flushed slopes, grazing often maintains the community and prevents progression of the community to scrub or woodland. Drainage and sward improvement have probably destroyed many smaller stands of this community and contributed to its very local distribution.

M26

Molinia caerulea and *Carex nigra* both abundant, with locally prominent *C. appropinquata* or *C. rostrata*. *Sanguisorba officinalis*, *Angelica sylvestris*, *Serratula tinctoria*, *Galium palustre* and *G. uliginosum* all frequent; bryophytes patchy.

Carex nigra often subordinate to *Molinia caerulea* in more grassy or rushy vegetation, with frequent and abundant *Festuca rubra*, *F. ovina*, *Holcus lanatus*, *Briza media*, *Deschampsia cespitosa*, *Anthoxanthum odoratum*, *Juncus acutiflorus* and *J. conglomeratus*; associates listed opposite all occasional.

M26a

Sanguisorba officinalis sub-community

This is generally the less species-rich sub-community. *Molinia caerulea* and *Carex nigra* are usually the most abundant plants, with one or both dominant in a swamp vegetation with large sedges as above. Smaller sedges can also occur, *Carex panicea* and *C. pulicaris* being common. Apart from *Agrostis stolonifera*, grasses are poorly represented and among small herbs only *Potentilla erecta*, *Ranunculus acris* and *Anemone nemorosa* occur with any frequency. The most common tall herbs are listed above. Bryophytes are patchy but better-developed than in M26b. *Calliergon cuspidatum*, *Lophocolea bidentata s.l.*, *Thuidium tamariscinum* and *Campylium stellatum* are frequent, with *Ctenidium molluscum*, *Plagiochila asplenioides*, *Campylium elodes* and *Aulacomnium palustre* being preferential.

This sub-community is found most extensively at Sunbiggin and Malham Tarns.

M26b

Festuca rubra sub-community

This sub-community appears more grassy and is usually developed on drier soils. *Molinia caerulea* is the usual dominant, forming the bulk of a rough sward together with the grasses listed above. Sedges are also common; *Carex nigra* and *C. panicea* can both show high cover and *C. flacca* and *C. pulicaris* can be frequent. Rushes are also common; *Juncus acutiflorus* and *J. conglomeratus* are preferential and *J. articulatus* also occurs. Taller dicotyledons remain frequent, although *Sanguisorba officinalis*, *Angelica sylvestris* and *Serratula tinctoria* are all scarce. *Geum rivale*, *Centaurea nigra* and *Leontodon hispidus* are more common than in M26a. In shorter swards, *Prunella vulgaris*, *Plantago lanceolata* and *Trifolium repens* can be found. Bryophytes are often poorly represented, but *Calliergon cuspidatum*, *Pseudoscleropodium purum* and *Lophocolea bidentata* remain frequent.

This sub-community has a scattered distribution through the dales along the upland fringes.

57

M27 *Filipendula ulmaria* – *Angelica sylvestris* mire

Although *Filipendula ulmaria* is frequent and locally abundant in a variety of vegetation types, in this community it forms the overwhelming dominant and the only constant. The dominants of other communities in which it occurs, tall helophytes, bulky sedges, rushes and rank grasses are, if present, all subordinate in this community. In the deep shade cast by *Filipendula* only scattered individuals or dispersed clumps of other species are found. The commonest accompanying tall herbs are *Angelica sylvestris*, *Valeriana officinalis* and *Rumex acetosa*. In the *Valeriana – Rumex* sub-community they are often accompanied by *Lychnis flos-cuculi*, *Succisa pratensis*, *Geum rivale* and sprawling *Galium palustre*. In the *Urtica – Vicia* sub-community these species are more scarce and *Urtica dioica* is very common with *Cirsium arvense*, *Epilobium hirsutum*, *Eupatorium cannabinum* and *Vicia cracca*. At low frequency throughout there can be scattered *Lythrum salicaria*, *Rumex crispus*, *R. sanguineus*, *Epilobium palustre*, *Equisetum palustre*, *E. arvense* and *E. fluviatile*.

There are few bulky monocotyledons; *Phragmites australis* can be common and *Phalaris arundinacea* is found occasionally. Rushes are few with *Juncus effusus* the most common. *Molinia caerulea* can also be found occasionally. Among smaller dicotyledons there can be *Ranunculus repens*, *Mentha aquatica*, *Lotus uliginosus*, and *Caltha palustris* with more occasional *Ranunculus acris*, *Cardamine pratensis*, *C. flexuosa*, *Potentilla anserina*, and *Polygonum hydropiper*. Bryophytes are few in number and of low cover.

This community is typically found where moist, reasonably rich, circumneutral soils occur in situations protected from grazing. It can be found in both topogenous and soligenous mires and is especially typical of silted margins of slow-moving streams and soakways, the edges of flushes and damp hollows, and also of artificial habitats such as along dykes and roadside ditches and around ponds. This community occurs throughout lowland Britain.

Both draining and grazing have reduced the extent of this community to small remnants in many places. The community cannot tolerate any other than very light or sporadic grazing and so stands often only persist outside enclosures, and around unreclaimed mires and flushes. For example, this community can be found in wet field bottoms and edges that have been fenced off, and alongside streams and ditches between pasture and boundaries. Progression to woodland, even in the absence of treatments such as grazing or mowing, appears to be slow.

M27

Associates listed for M27a at most occasional, but sparse *Phragmites australis* common, often with prominent clumps of *Urtica dioica*, *Eupatorium cannabinum* and *Epilobium hirsutum*. The sprawling herbs *Galium aparine* and *Vicia cracca* are common.

Associates listed opposite at most occasional but *Juncus effusus* and *Holcus lanatus* are constant, and *Juncus acutiflorus* and *Molinia caerulea* occasional in ranker swards with *Anthoxanthum odoratum*, *Agrostis stolonifera*, *Mentha aquatica* and *Lotus uliginosus* quite common.

Angelica sylvestris, *Valeriana officinalis*, *Rumex acetosa*, *Lychnis flos-cuculi*, *Succisa pratensis* and *Geum rivale* are common among taller associates, with *Caltha palustris*, *Ranunculus flammula*, *R. repens*, *R. acris*, *Cardamine flexuosa* and *C. pratensis* occasional to frequent below and *Galium palustre*, *G. uliginosum* and *Lathyrus pratensis* climbing or sprawling.

M27a

Valeriana officinalis – *Rumex acetosa* sub-community

Filipendula ulmaria is abundant and dominant. The vegetation may be species-poor, but overall it is characterised by a variety of associates. Most common are *Angelica sylvestris* and *Valeriana officinalis*, both sometimes locally abundant, with *Rumex acetosa*, *Lychnis flos-cuculi*, *Succisa pratensis* and *Geum rivale* are less common. Among smaller herbs can be the species listed above with *Stellaria alsine* and *Ajuga reptans* and *Galium palustre* as the most common sprawling plant. Apart from *Poa trivialis*, preferential here, grasses and rushes are infrequent. *Carex rostrata* is quite common and can occur with *Menyanthes trifoliata* and *Potentilla palustris*. Bryophytes are better developed in this sub-community, with *Brachythecium rutabulum* the most common species.

This sub-community is the usual form in northern England and in southern and eastern Scotland.

M27b

Urtica dioica – *Vicia cracca* sub-community

Filipendula ulmaria and a variety of tall herbs again provide the main structural element. *Angelica sylvestris*, *Cirsium palustre* and *Lythrum salicaria* are occasional, but others present in M27a, e.g. *Valeriana officinalis* and *Rumex acetosa*, become scarce or absent. *Urtica dioica*, however, is very common and is found with occasional *Eupatorium cannabinum* and *Epilobium hirsutum* forming patches. Scattered throughout can be *Cirsium arvense* and *Centaurea nigra*. *Phragmites australis* can be common; alternatively there may be tussocks of *Arrhenatherum elatius* and some *Holcus lanatus* or scattered clumps of rushes. Smaller herbs and bryophytes are few and sparse.

This sub-community is found in central, southern and eastern Britain.

M27c

Juncus effusus – *Holcus lanatus* sub-community

Filipendula ulmaria is still the most abundant species, but other tall herbs such as *Angelica sylvestris*, *Valeriana officinalis*, *Cirsium palustre* and *Rumex acetosa* occur occasionally. Of greater importance, rushes and grasses may have moderate abundance. *Juncus effusus* and *Holcus lanatus* are both constant and *J. acutiflorus* and *Molinia caerulea* occasional, with a range of other grasses. In some stands *Mentha aquatica* and *Lotus uliginosus* are frequent. In others *Iris pseudacorus* and/or *Oenanthe crocata* can be prominent.

The sub-community is western in distribution.

M28 *Iris pseudacorus* – *Filipendula ulmaria* mire

In its typical form this is a luxuriant and species-rich community with *Iris pseudacorus* more abundant than *Oenanthe crocata*, although both are constants, except in the far north of Scotland where *O. crocata* is not found. Other tall herbs are nearly always present, though only *Lycopus europaeus*, *Rumex crispus* and *Scutellaria galericulata* are frequent throughout the community. Other species such as *Rumex acetosa*, *Lychnis flos-cuculi*, *Angelica sylvestris*, *Valeriana officinalis*, *Cirsium palustre*, *C. arvense* and *Urtica dioica* are often common and conspicuous but preferential to particular sub-communities.

Rushes and grasses are frequently important. *Juncus effusus* and *J. acutiflorus* are common, as are *Poa trivialis* and *Agrostis stolonifera*. There are a variety of smaller dicotyledons. Some typically occur as scattered plants, for example *Ranunculus acris*, *Caltha palustris*, *Stellaria alsine*, *Mentha aquatica* and *Hydrocotyle vulgaris*, while *Ranunculus repens* and *Potentilla anserina* form mats with high local cover. On patches of wet and open ground, annuals may be prolific, such as *Polygonum hydropiper*, *Montia fontana*, and on cattle-poached mud, *Ranunculus sceleratus*. On salt-marsh transitions *Atriplex prostrata* and *Matricaria maritima* may be frequent with *Samolus valerandi*, *Oenanthe lachenalii* and halophytic herbs. Bryophytes are few, with *Eurhynchium praelongum* being the commonest throughout.

This community is confined to moist, more nutrient-rich soils along the oceanic seaboard of Britain. It is especially characteristic of the freshwater seepage zone along the upper edge of saltmarshes in the sheltered sea-lochs of western Scotland. Other situations in which it occurs are over stabilised shingle down the west coast and in wetter hollows and flushes on raised beach platforms and gentle cliff slopes. The community is the oceanic counterpart of the *Filipendula ulmaria – Angelica sylvestris* mire (M27) and is largely confined to the west coast of Britain. In particular it is found in west Scotland from Orkney and Shetland southwards, with scattered stands in south-west England and west Wales.

The community was probably once much more widespread in south-west England and west Wales but it may have been largely destroyed in its salt-marsh habitat by human interference with the transitional upper zones. The community, where it does occur, is often not heavily grazed and it appears to be a relatively stable vegetation type with only a slow progression to scrub or woodland.

M28

Juncus effusus and/or J. acutiflorus constant and patchily abundant with frequent Rumex acetosa, Cirsium palustre, Epilobium palustre, Lychnis flos-cuculi, Ranunculus acris, Caltha palustris, Lotus uliginosus and Galium palustre.

Species listed opposite occasional at most, but Urtica dioica and Cirsium arvense constant with Galium aparine and occasional to frequent Elymus repens, Stellaria media, Arrhenatherum elatius and Dactylis glomerata.

Groups of species opposite, and even Filipendula ulmaria, infrequent in rather open vegetation with Atriplex prostrata and Samolus valerandi common and sporadic records for maritime plants.

M28a

Juncus spp. sub-community

This is the richest sub-community in which other dicotyledons, rushes and grasses form a consistent associated flora. Iris pseudacorus is generally a clear dominant although both Oenanthe crocata and Filipendula ulmaria can be patchily abundant with the above Juncus spp. Among the taller herbs, Lycopus europaeus, Rumex crispus, Scutellaria galericulata, Angelica sylvestris and Valeriana officinalis, all occur frequently to occasionally. Common grasses are Festuca rubra, Holcus lanatus, Anthoxanthum odoratum, Poa pratensis and Elymus repens forming scattered tussocks, and with Carex otrubae and trailing Galium palustre.

M28b

Urtica dioica – Galium aparine sub-community

The vegetation here, although as tall and luxuriant as in M28a, is less species-rich. Iris pseudacorus is still dominant and both Oenanthe crocata and Filipendula ulmaria remain frequent, but apart from Lycopus europaeus and Scutellaria galericulata, the only other common taller dicotyledons are Urtica dioica and Cirsium arvense which can be abundant. Other taller species of M28a are occasional or scarce and Galium aparine replaces G. palustre. Grasses are often conspicuous with Poa trivialis and Agrostis stolonifera very common as patchy carpets interspersed with the other species listed above.

M28c

Atriplex prostrata – Samolus valerandi sub-community

Although Iris pseudacorus and Oenanthe crocata can be abundant, taller associates are generally lacking; even Filipendula ulmaria is scarce and Lycopus europaeus and Rumex crispus are only present as scattered plants. Among smaller plants the commonest grasses are Agrostis stolonifera and Festuca rubra. The most common preferentials are plants tolerant of saline habitats, including Atriplex prostrata, the most common, with Samolus valerandi, Oenanthe lachenalii, Matricaria maritima, Triglochin maritima and Glaux maritima.

M29 *Hypericum elodes – Potamogeton polygonifolius* soakway

This community has a very distinctive appearance, typically consisting of low creeping or floating mats of *Hypericum elodes* and *Potamogeton polygonifolius*. Very often, unless the ground has been badly trampled by grazing animals, these two constants are set in a carpet of submerged *Sphagnum auriculatum*, sometimes with *S. cuspidatum*, *S. palustre* or *S. recurvum*. Other bryophytes are sparse but *Polytrichum commune* or *Aulacomnium palustre*, and *Drepanocladus exannulatus*, *D. revolvens*, and *Calliergon cuspidatum*, may be locally abundant.

Other vascular plants are scattered. *Juncus bulbosus* and *Ranunculus flammula* are the only constants but *Hydrocotyle vulgaris*, *Anagallis tenella*, *Drosera rotundifolia*, *Narthecium ossifragum*, and *Galium palustre* can all be moderately frequent, along with sedges such as *Carex demissa*, *C. echinata*, *C. panicea* and *C. nigra*. There can also be sparse shoots or small patches of *Molinia caerulea*, *Agrostis canina* ssp. *canina*, *Juncus articulatus*, *J. effusus*, *J. acutiflorus*, *Eleocharis multicaulis*, *Eriophorum angustifolium* and *Rhynchospora alba*. *Carex rostrata* can also be found in some stands. Two rare species associated with this community are *Galium debile*, in the New Forest, and the fern *Pilularia globulifera*.

This community is characteristic of shallow soakways and pools in peats and peaty mineral soils with fluctuating water levels, such as seepages and runnels around mires and in heathland pools, at moderate altitudes. The water is typically clear, still or gently-flowing, moderately acid to neutral, with a pH between 4 and 5.5, and probably quite oligotrophic. This vegetation is confined to the warm oceanic parts of Britain and extends in a well-defined zone from west Surrey through the New Forest to the South-West Peninsula and north through Wales to Galloway. It may well be found further north, following the distribution of *Hypericum elodes*.

This soakway appears to be a stable vegetation type in the absence of nutrient enrichment. In situations where there is some nutrient enrichment, grazing and trampling may help continually set back any tendency to succession. Trampling by grazing animals can also play a part in keeping the vegetation open and varied, although heavy poaching can be deleterious to the *Sphagnum* carpet.

No sub-communities.

M30 Related vegetation of seasonally-inundated habitats

Other vegetation of the same type as *Hypericum elodes – Potamogeton polygonifolius* soakway (M29), and characteristic of similar, seasonally inundated habitats, with rather base-poor and only moderately enriched waters, has only been poorly sampled. Some examples, lacking *Hypericum elodes* but otherwise the same as M29, may be regarded as impoverished stands of that community, though it must be noted that species such as *Potamogeton polygonifolius*, *Eriophorum angustifolium*, *Juncus bulbosus/ kochii* and *Sphagnum auriculatum* also occur with some frequency in bog-pool and poor-fen vegetation.

There are also stands in which *Eleocharis multicaulis* is strongly dominant with little or no *Hypericum elodes* or *Potamogeton polygonifolius*; these look similar to the *Eleocharitetum multicaulis* recorded from Eire and elsewhere in western Europe. *Deschampsia setacea* is listed as a characteristic species of such vegetation, and in Britain this rare species is typical of this type of habitat. *Scirpus fluitans* can also be found dominating in swards which lack some of the typical plants of M29, and in the New Forest and Cornwall, *Baldellia ranunculoides* is a frequent and conspicuous component of low-growing vegetation in seasonally wet pools.

All these vegetation types, along with M29, have been grouped in the Hydrocotylo-Baldellion alliance, comprising assemblages of mesotrophic to oligotrophic, and periodically fluctuating waters.

No sub-communities.

M31 *Anthelia julacea – Sphagnum auriculatum* spring

In this community *Anthelia julacea* forms cushions up to a metre or more in thickness and several square metres in extent. Its associated flora is species-poor and vascular plants are sparse. Among other bryophytes, *Sphagnum auriculatum*, *Marsupella marginata* and *Scapania undulata* are all constant, the first of which can form prominent patches. *Racomitrium lanuginosum* and *Philonotis fontana* are also frequent with occasional *Calliergon sarmentosum*, *Campylopus atrovirens*, *Polytrichum commune* and *Racomitrium fasciculare*. Rare bryophytes which have been recorded in this community include *Anthelia juratzkana*, which almost totally replaces *A. julacea* in some stands, and *Pohlia ludwigii*.

The commonest vascular plant is *Deschampsia cespitosa*. Scattered plants of *Nardus stricta* occur quite often with occasional *Narthecium ossifragum*, *Pinguicula vulgaris*, *Carex demissa*, and *Saxifraga stellaris*. Less frequent are *Eriophorum angustifolium*, *Carex bigelowii*, *C. nigra*, *Festuca vivipara*, *Agrostis canina*, *A. stolonifera*, *Juncus bulbosus*, *Thalictrum alpinum*, and *Viola palustris*.

This is a montane community typical of often skeletal mineral and organic soils kept more or less permanently wet by the trickling of acid and oligotrophic waters, of pH 4.5-5.0, frequently derived at higher altitudes from snow-melt. It occurs at moderate to high altitudes, from about 400 m to over 1000 m, marking out areas of less vigorous seepage among tracts of montane grasslands and grass-heaths or among snow-bed vegetation. It is a local but widespread community found through much of Scotland, the Lake District and Snowdonia.

Anthelia julacea – Sphagnum auriculatum spring is an essentially stable community in the harsh environment in which it characteristically occurs.

No sub-communities.

M32 *Philonotis fontana –*
Saxifraga stellaris spring

These bryophyte-dominated springs, flushes and rills are striking in appearance. *Philonotis fontana* is usually dominant and obvious by its fresh-green colour. The golden-green *Dicranella palustris* and *Scapania undulata* are often also abundant. These species together often form cushions or plush mats. Less consistent but sometimes prominent are *Sphagnum auriculatum*, *Scapania uliginosa*, *Calliergon sarmentosum*, *Drepanocladus exannulatus*, *D. fluitans* and the characteristic, but not very frequent, *Jungermannia exsertifolia*. *Bryum pseudotriquetrum* is also common, but not as consistent as in more base-rich springs, and *Cratoneuron commutatum* and *C. filicinum* are scarce.

The associated vascular flora is rather varied in composition and cover and typically species-poor. The only constant is *Saxifraga stellaris*, but *Deschampsia cespitosa* (with ssp. *alpina* at higher altitudes) is often found in small quantities with scattered *Stellaria alsine*. More occasional are *Festuca rubra*, *Anthoxanthum odoratum*, *Agrostis stolonifera* and *A. canina*, all usually at low cover, with sparse *Viola palustris*, *Nardus stricta*, *Carex bigelowii* and *C. panicea*. In stands which are perhaps less base-poor, *Montia fontana* and *Chrysosplenium oppositifolium* become frequent and abundant, along with *Caltha palustris*, *Cardamine pratensis* and a number of other associated species.

This is a community of springs and rills at moderate to high altitudes, above 450 m to over 1000 m, where there is irrigation with circum-neutral and oligotrophic waters with a pH of around 4.5-6. This is one of the most common and widespread types of spring vegetation in the uplands of north-west Britain and is dependent on sustained and vigorous irrigation by ground-waters, common in the wetter parts of the country. It marks out permanent springs of a well-defined character, also diffuse flushes and seep-age lines, rills and small streams and occasionally steep, dripping ground. In some places snow-melt may be an important water source. It is found on a variety of waterlogged soils from fragmentary accumulation of silt among stones to flushed peats and gleys. The community is common through the Scottish Highlands, the Southern Uplands, the Lake District and north Wales and over the non-calcareous parts of the Pennines. It can also be found as fragmentary stands at lower altitudes without the montane element in the vegetation, particularly at the southern limit of the range of this community.

The harsh montane environment has a striking influence on the composition of the community and though stands can be grazed and trampled, climatic and soil conditions probably play the major part in maintaining the vegetation as an effective climax. At lower altitudes, it could perhaps show some successional development in ungrazed situations.

```
                          ┌──────────┐
                          │   M32    │
                          └──────────┘
        ┌────────────────────┴──────────────────────┐
```

Sphagnum auriculatum often abundant, with *Calliergon sarmentosum* and *Scapania uliginosa* occasional and locally prominent. *Montia fontana* and *Chrysosplenium oppositifolium* very occasional.	*Sphagnum auriculatum* and other bryophytes listed opposite very scarce but *Bryum pseudotriquetrum* frequent. *Montia fontana* and *Chrysosplenium oppositifolium* constant and sometimes abundant.

M32a

Sphagnum auriculatum sub-community

In this species-poor sub-community the bryophyte mat consists of mixtures of *Philonotis fontana, Scapania undulata* and *Dicranella palustris* with *Sphagnum auriculatum* strongly preferential and often abundant. *Calliergon sarmentosum* and *Scapania uliginosa* are occasional and *Polytrichum commune* and *Hygrohypnum ochraceum* occur at low frequencies. Among the vascular plants only *Saxifraga stellaris* and *Deschampsia cespitosa* are constant but other grasses such as *Agrostis stolonifera, A. capillaris, Anthoxanthum odoratum* and *Festuca rubra* are quite frequent as scattered tufts.

This sub-community occurs mainly on the harder acidic quartzites and sandstones of the north-west Highlands.

M32b

Montia fontana – Chrysosplenium oppositifolium sub-community

Vascular plants are more numerous and varied, although bryophytes still generally have dominance. *Philonotis fontana, Dicranella palustris* and *Scapania undulata* are all very common and each, especially the first, can be abundant. *Bryum pseudotriquetrum* is frequent and *Jungermannia exsertifolia* occasional. Among the vascular plants, *Montia fontana* and *Chrysosplenium oppositifolium* have high frequencies with *Saxifraga stellaris*, and are sometimes abundant. Along with *Stellaria alsine* there are often very small plants of *Caltha palustris* (ssp. *minor*) and *Cardamine pratensis*. *Epilobium palustre* can sometimes be found but more distinctive are the frequent occurrence of *E. alsinifolium* and *E. anagallidifolium*. Grasses can be quite common; *Deschampsia cespitosa* is often joined by *Anthoxanthum odoratum, Agrostis canina* and several other grasses together with some sedges.

This community is associated with a range of substrates slightly more base-rich than those of M32a.

M33 *Pohlia wahlenbergii* var. *glacialis* spring

In this community *Pohlia wahlenbergii* var. *glacialis* dominates in spongy carpets of a bright apple-green colour, often of small extent, but exceptionally up to 200 m². Few other bryophytes occur with any frequency although *P. ludwigii* is a constant. *Philonotis fontana* can be prominent, although not with the high cover found in *Philonotis fontana – Saxifraga stellaris* spring (M32). Other bryophytes recorded occasionally are *Hygrohypnum luridum*, *Bryum weigelii*, *Calliergon stramineum*, *Scapania undulata*, *S. uliginosa*, *Dicranella palustris* and *Marchantia alpestris*.

In this carpet there are only a few vascular plants. *Deschampsia cespitosa* (presumably ssp. *alpina*) and *Saxifraga stellaris* are constant, but *Cerastium cerastoides* is quite often found and there can be *Stellaria alsine*, *Chrysosplenium oppositifolium*, *Epilobium anagallidifolium*, *Veronica serpyllifolia* var. *humifusa*, and *Rumex acetosa*. Other rare plants found occasionally are *Epilobium alsinifolium*, *Alopecurus alpinus* and *Phleum alpinum*.

This community is strictly confined to spring-heads associated with late snow-beds where there is vigorous irrigation by cold waters. The flushing waters, and often sloppy, ill-structured mixtures of mineral and organic matter beneath the moss carpet, are base-poor and oligotrophic. Although *Pohlia wahlenbergii* var. *glacialis* occurs over quite a range of altitudes through the uplands of Wales, Cumbria and Scotland, it is only found with the kind of dominance characteristic here within the high montane zone at altitudes generally above 850 m. Within this area, which includes the central and north-western Highlands of Scotland, the community is further restricted to situations where the snow lies longest, especially on north- and east-facing slopes.

The general climatic and edaphic features determine the overall character of this community with its cold-tolerant plants and montane species.

No sub-communities.

M34 *Carex demissa – Koenigia islandica* flush

This is an open vegetation type with a bryophyte-dominated carpet broken by areas of wet, silty and stony ground. *Scapania undulata*, *Calliergon sarmentosum* and *Blindia acuta* are all common and each can be abundant, with occasional patches of *Dicranella palustris*, *Philonotis fontana*, *Drepanocladus revolvens*, *Marsupella aquatica* and *Sphagnum auriculatum*. Scattered through this and over the rills are plants of *Carex demissa*, *Koenigia islandica*, *Deschampsia cespitosa*, *Saxifraga stellaris*, *Juncus triglumis*, *J. bulbosus* and the rare *J. biglumis* and *Sagina saginoides*. All these are generally of low cover, though many can show a measure of abundance and *Koenigia*, although individual plants are small, can cover quite a large ground area.

This community occurs on ground which is kept periodically moist by circumneutral and oligotrophic waters. Typically it is found in open silty or stony flushes fed by vigorous seepage from springs issuing at moderately high altitudes, over 500 m, from basalt. In their base status, with pH values around 6.0, and their low cation content, the waters are similar to those which feed the *Philonotis fontana – Saxifraga stellaris* spring community (M32), which often occupies the spring-heads above the flushes. The community is confined to Skye, where it occurs scattered along the Trotternish Ridge extending several kilometres north of the Storr. It forms one of several communities in which *Koenigia* can be found.

No sub-communities.

M35 *Ranunculus omiophyllus –* *Montia fontana* rill

These rills typically have a rather crowded, though not always continuous, cover of vascular plants and bryophytes. Much of the growth is often submerged in the shallow waters, with a floating or shortly emergent canopy. *Ranunculus omiophyllus* is often abundant, frequently with *Montia fontana*. Floating leaves of *Potamogeton polygonifolius* are commonly prominent and there can be local patches of *Agrostis stolonifera, Glyceria fluitans, Juncus bulbosus, J. articulatus* and *Callitriche stagnalis*, with scattered plants of *Ranunculus flammula*, a constant, *Myosotis secunda* and *Stellaria alsine*. More occasional are *Ranunculus repens, Equisetum palustre, Hydrocotyle vulgaris, Galium palustre* and *Lotus uliginosus. Juncus bufonius* and *Scirpus setaceus* can sometimes be seen on open mud.

Bryophytes can contribute substantially to the cover although there are only a few frequent species. *Sphagnum auriculatum* is a constant and often grows semi-submerged with patches of *Philonotis fontana* but, apart from occasional *Polytrichum commune*, other species are sparse, with only occasional records of *Calliergon cuspidatum, C. stramineum, Drepanocladus exannulatus, D. vernicosus, Scapania irrigua* and *Rhytidiadelphus squarrosus.*

This community is typical of spring-heads and rills at moderate altitudes in south-western Britain, where there is irrigation by circumneutral and probably quite oligotrophic waters. These are typically rather base- and nutrient-poor with pH values ranging from 4.5 to 6.5 over acidic rocks. It has been recorded only from south-western England, Wales, and around the Lake District. It may occur throughout the range of *R. omiophyllus* in Britain.

No sub-communities.

M36 Lowland springs and streambanks of shaded situations

There is a clear contrast, among the Cardamino – Montion springs and flushes of acid to circumneutral habitats, between the upland communities which have been described, where *Montia fontana, Saxifraga stellaris* and *Philonotis fontana* are conspicuous, and the vegetation of lowland and often shaded situations. In these, *Chrysosplenium oppositifolium* occurs with bryophytes such as *Hookeria lucens, Rhizomnium punctatum, Trichocolea tomentella, Pellia epiphylla* and *Conocephalum conicum*. This type of vegetation has not been separately sampled here but it figures in the field and ground layers of various wet woodlands, notably the *Alnus – Carex, Alnus – Urtica* and *Alnus – Fraxinus – Lysimachia* types, where it is distinctive of seepage lines and damp stream banks, quite often with *Cardamine flexuosa, C. amara* and *Chrysosplenium alternifolium*. Similar mixtures of plants can be found widely through lowland Britain, especially in the wetter west and around the upland fringes, along stream-sides and wet banks, probably once wooded, but where shade is now provided by tall herbs or by the aspect of the site. These need further sampling.

No sub-communities.

M37 *Cratoneuron commutatum – Festuca rubra* spring

Cratoneuron commutatum occurs frequently in a variety of calcareous mires, but here it is consistently dominant in large masses, often forming prominent mounds or banks. In some stands of the same general floristic composition, *C. filicinum* accompanies or totally replaces it. Other bryophytes can make a contribution, but typically a minor one. However, the constant *Bryum pseudotriquetrum* is very common. Occasional species include *Philonotis fontana*, *P. calcarea*, *Aneura pinguis*, *Pellia endiviifolia*, *Drepanocladus revolvens*, *Gymnostomum recurvirostrum*, *G. aeruginosum*, *Brachythecium rivulare* and *Dicranella palustris.* Very typically there is some tufa deposition allowing the mat to build into mounds. The vascular element is typically species-poor and of low total cover. There may be considerable variation in associated flora and, particularly where stands are developed on gently-sloping ground, a richer and more extensive layer can be found, coming close to the *Cratoneuron commutatum – Carex nigra* spring (M38). Often, however the only species present are *Festuca rubra*, *Cardamine pratensis* and *Saxifraga aizoides*, the last of which is absent from southern Scotland and Wales. Occasional herbs include *Agrostis stolonifera*, *Deschampsia cespitosa*, *Equisetum palustre*, *Chrysosplenium oppositifolium*, *Poa trivialis*, *Carex panicea*, *C. nigra* and *C. dioica.*

This is a community of ground kept permanently moist by irrigation with base-rich, calcareous and generally oligotrophic waters. It is dependent on sustained irrigation common in areas of higher rainfall. Here it can be found marking out spring-heads, seepage lines and drip zones in areas of lime-rich bedrocks, where waters emerge along bedding planes or at junctions with impervious substrates. Provided the ground is permanently wet, the community can even occur on vertical surfaces and bare rock, forming curtain-like masses. The community can be found throughout the north-western uplands of Britain with its more Arctic-Alpine element best developed in the Scottish Highlands, with outliers in the Lake District and Upper Teesdale. Springs dominated by *Cratoneuron* species also occur widely, but locally, in the British lowlands, and further sampling of these is needed.

In most circumstances it is a permanent community maintained by edaphic and climatic conditions of the environment. On gentle slopes, trampling by grazing stock or deer often plays an important part in maintaining the characteristically open conditions of flushed soils, but trampling and grazing can have an adverse effect on the bryophyte carpet.

No sub-communities.

M38 *Cratoneuron commutatum – Carex nigra* spring

This type of spring preserves the same pattern of dominance by *Cratoneuron commutatum* (again occasionally supplemented or replaced by *C. filicinum*) as in *Cratoneuron commutatum – Festuca rubra* spring (M37), but the associated flora is much richer. This is partly seen among the bryophytes. *Bryum pseudotriquetrum* and *Philonotis fontana* are the commonest and can have moderately high cover, and there are many others which can occur locally as prominent patches. These include calcicolous species such as *Aneura pinguis*, *Fissidens adianthoides*, *Philonotis calcarea*, *Ctenidium molluscum*, *Cinclidium stygium*, *Drepanocladus revolvens* and *Campylium stellatum*.

The increased richness is most seen among the vascular plants. Small sedges are noticeable. *Carex demissa*, *C. nigra* and *C. panicea* are constant and often abundant, and *C. pulicaris*, *C. flacca* and *C. dioica* are common. There are frequent scattered plants of *Cardamine pratensis*, *Selaginella selaginoides*, *Leontodon autumnalis*, *Polygonum viviparum*, *Trifolium repens*, *Cirsium palustre*, *Ranunculus flammula*, *Sagina nodosa*, *Juncus triglumis*, *J. articulatus*, *J. bulbosus*, *Cera-stium fontanum*, *Prunella vulgaris*, *Caltha palustris*, *Galium palustre*, *Equisetum palustre*, *Achillea ptarmica*, *Cochlearia officinalis* (often ssp. *alpina*), *Triglochin palustris*, *Ranunculus acris*, *Anthoxanthum odoratum*, *Festuca ovina*, *Epilobium anagallidifolium* and, in north England, the introduced *E. nerteroides*. In Teesdale this community is the locus for *Saxifraga hirculus*.

This vegetation is confined to montane springs and flushes strongly irrigated by base-rich, calcareous and oligotrophic waters. As in M37, sites of sustained irrigation with waters draining from lime-rich bedrock are marked out and tufa encrustation is often seen. It is very local around Upper Teesdale in the north Pennines and in the central Highlands of Scotland, mostly above 650 m altitude.

Although the harsh climatic and edaphic conditions exert a strong influence on the structure and composition of the vegetation, heavy grazing plays a major role in maintaining the distinctive richness of the community, and it is this trampling and cropping by sheep and deer which is responsible for the most obvious floristic differences between this community and M37.

No sub-communities.

4 Dendrogram keys to heath communities

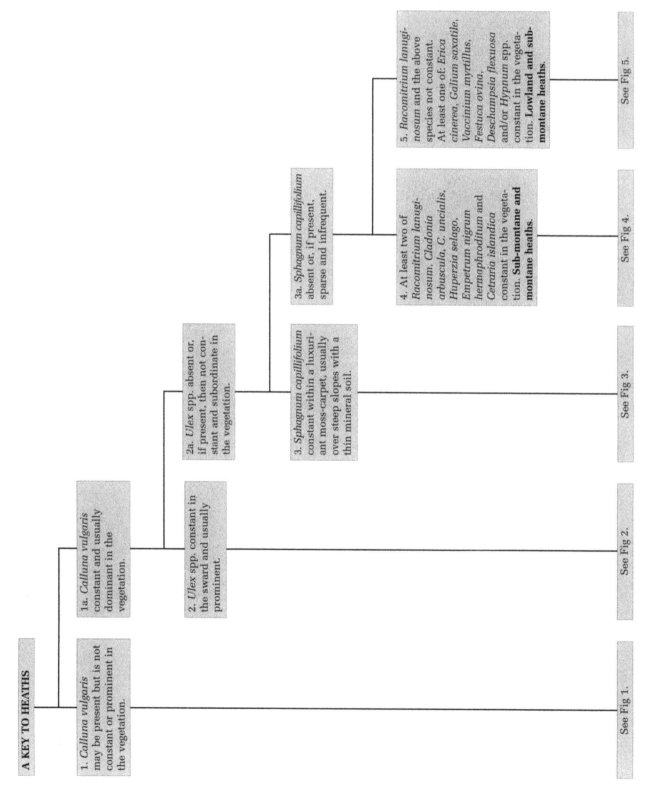

A KEY TO HEATHS

1. *Calluna vulgaris* may be present but is not constant or prominent in the vegetation.

1a. *Calluna vulgaris* constant and usually dominant in the vegetation.

2. *Ulex* spp. constant in the sward and usually prominent.

2a. *Ulex* spp. absent or, if present, then not constant and subordinate in the vegetation.

3. *Sphagnum capillifolium* constant within a luxuriant moss-carpet, usually over steep slopes with a thin mineral soil.

3a. *Sphagnum capillifolium* absent or, if present, sparse and infrequent.

4. At least two of *Racomitrium lanuginosum*, *Cladonia arbuscula*, *C. uncialis*, *Huperzia selago*, *Empetrum nigrum hermaphroditum* and *Cetraria islandica* constant in the vegetation. **Sub-montane and montane heaths.**

5. *Racomitrium lanuginosum* and the above species not constant. At least one of: *Erica cinerea*, *Galium saxatile*, *Vaccinium myrtillus*, *Festuca ovina*, *Deschampsia flexuosa* and/or *Hypnum* spp. constant in the vegetation. **Lowland and sub-montane heaths.**

See Fig 1.

See Fig 2.

See Fig 3.

See Fig 4.

See Fig 5.

Heaths Figure 1

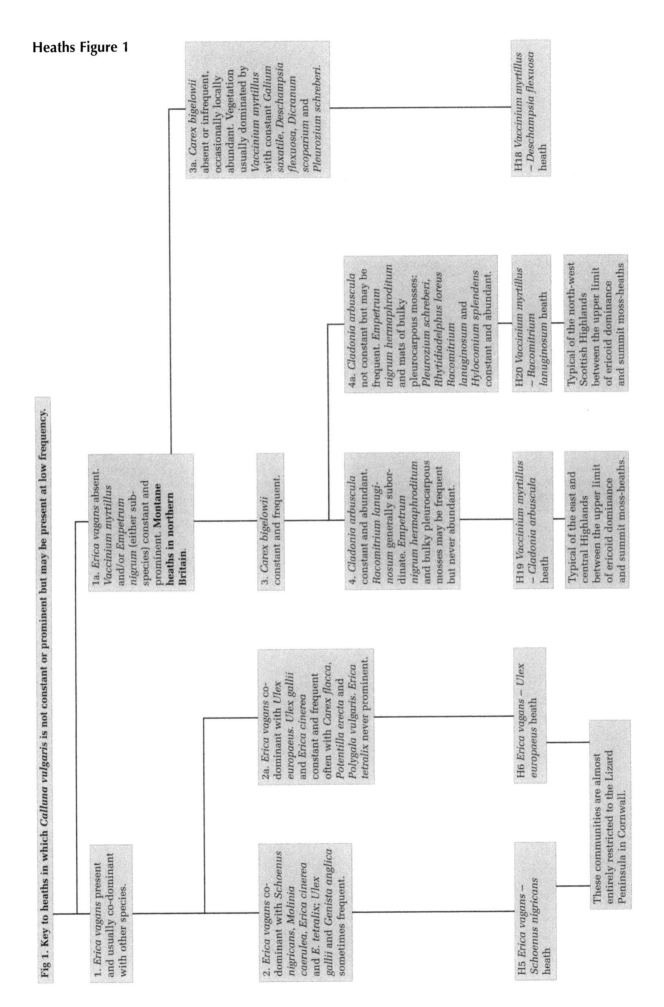

Fig 1. Key to heaths in which *Calluna vulgaris* is not constant or prominent but may be present at low frequency.

1. *Erica vagans* present and usually co-dominant with other species.

1a. *Erica vagans* absent. *Vaccinium myrtillus* and/or *Empetrum nigrum* (either subspecies) constant and prominent. **Montane heaths in northern Britain.**

2. *Erica vagans* co-dominant with *Schoenus nigricans, Molinia caerulea, Erica cinerea* and *E. tetralix; Ulex gallii* and *Genista anglica* sometimes frequent.

2a. *Erica vagans* co-dominant with *Ulex europaeus. Ulex gallii* and *Erica cinerea* constant and frequent often with *Carex flacca, Potentilla erecta* and *Polygala vulgaris. Erica tetralix* never prominent.

3. *Carex bigelowii* constant and frequent.

3a. *Carex bigelowii* absent or infrequent, occasionally locally abundant. Vegetation usually dominated by *Vaccinium myrtillus* with constant *Galium saxatile, Deschampsia flexuosa, Dicranum scoparium* and *Pleurozium schreberi.*

4. *Cladonia arbuscula* constant and abundant. *Racomitrium lanuginosum* generally subordinate. *Empetrum nigrum hermaphroditum* and bulky pleurocarpous mosses may be frequent but never abundant.

4a. *Cladonia arbuscula* not constant but may be frequent. *Empetrum nigrum hermaphroditum* and mats of bulky pleurocarpous mosses: *Pleurozium schreberi, Rhytidiadelphus loreus Racomitrium lanuginosum* and *Hylocomium splendens* constant and abundant.

H5 *Erica vagans – Schoenus nigricans* heath

H6 *Erica vagans – Ulex europaeus* heath

H18 *Vaccinium myrtillus – Deschampsia flexuosa* heath

H19 *Vaccinium myrtillus – Cladonia arbuscula* heath

H20 *Vaccinium myrtillus – Racomitrium lanuginosum* heath

These communities are almost entirely restricted to the Lizard Peninsula in Cornwall.

Typical of the east and central Highlands between the upper limit of ericoid dominance and summit moss-heaths.

Typical of the north-west Scottish Highlands between the upper limit of ericoid dominance and summit moss-heaths

Heaths Figure 2

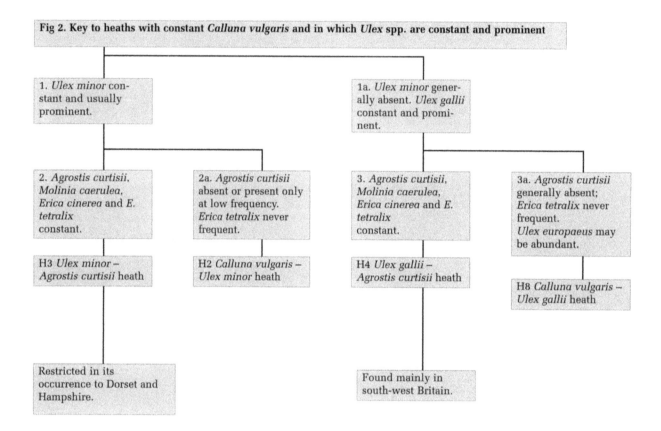

Fig 2. Key to heaths with constant *Calluna vulgaris* and in which *Ulex* spp. are constant and prominent

1. *Ulex minor* constant and usually prominent.

1a. *Ulex minor* generally absent. *Ulex gallii* constant and prominent.

2. *Agrostis curtisii*, *Molinia caerulea*, *Erica cinerea* and *E. tetralix* constant.

2a. *Agrostis curtisii* absent or present only at low frequency. *Erica tetralix* never frequent.

3. *Agrostis curtisii*, *Molinia caerulea*, *Erica cinerea* and *E. tetralix* constant.

3a. *Agrostis curtisii* generally absent; *Erica tetralix* never frequent. *Ulex europaeus* may be abundant.

H3 *Ulex minor – Agrostis curtisii* heath

H2 *Calluna vulgaris – Ulex minor* heath

H4 *Ulex gallii – Agrostis curtisii* heath

H8 *Calluna vulgaris – Ulex gallii* heath

Restricted in its occurrence to Dorset and Hampshire.

Found mainly in south-west Britain.

Heaths Figure 3

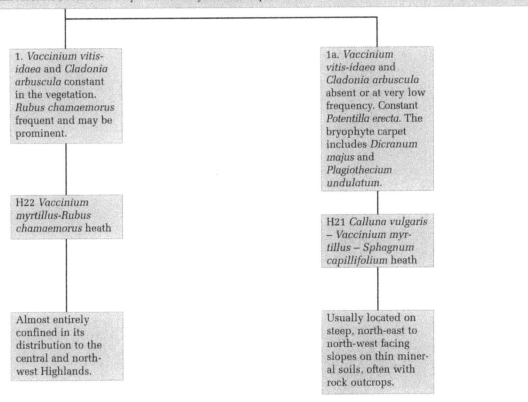

Fig 3. Key to heaths with constant *Calluna vulgaris* and *Sphagnum capillifolium* over a luxuriant bryophyte-rich carpet. Constants include: *Vaccinium myrtillus, Empetrum nigrum nigrum, Deschampsia flexuosa, Rhytidiadelphus loreus, Pleurozium schreberi, Dicranum scoparium* and *Hylocomium splendens*.

1. *Vaccinium vitis-idaea* and *Cladonia arbuscula* constant in the vegetation. *Rubus chamaemorus* frequent and may be prominent.

1a. *Vaccinium vitis-idaea* and *Cladonia arbuscula* absent or at very low frequency. Constant *Potentilla erecta*. The bryophyte carpet includes *Dicranum majus* and *Plagiothecium undulatum*.

H22 *Vaccinium myrtillus-Rubus chamaemorus* heath

H21 *Calluna vulgaris – Vaccinium myrtillus – Sphagnum capillifolium* heath

Almost entirely confined in its distribution to the central and north-west Highlands.

Usually located on steep, north-east to north-west facing slopes on thin mineral soils, often with rock outcrops.

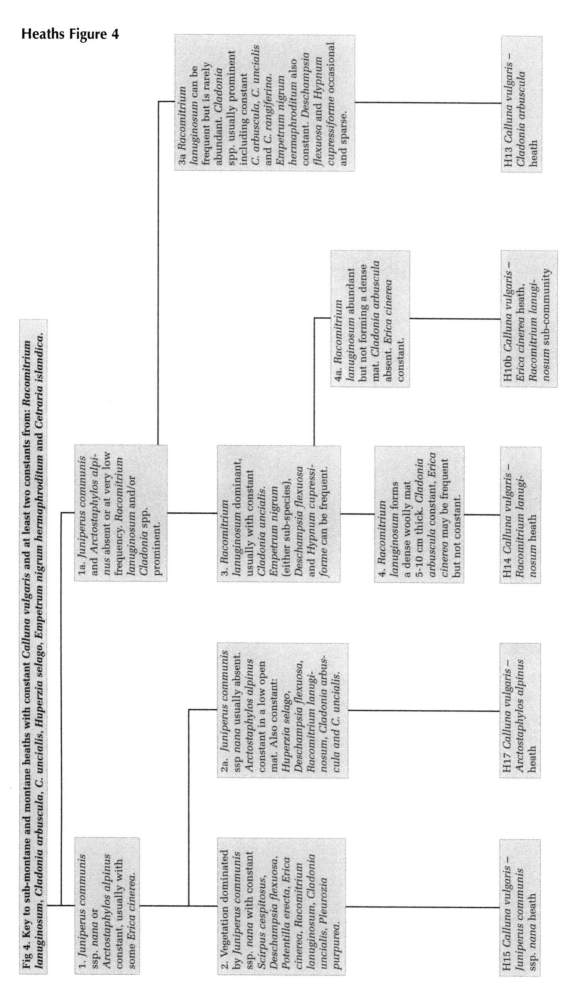

Fig 4. Key to sub-montane and montane heaths with constant *Calluna vulgaris* and at least two constants from: *Racomitrium lanuginosum, Cladonia arbuscula, C. uncialis, Huperzia selago, Empetrum nigrum hermaphroditum* and *Cetraria islandica*.

Heaths Figure 5

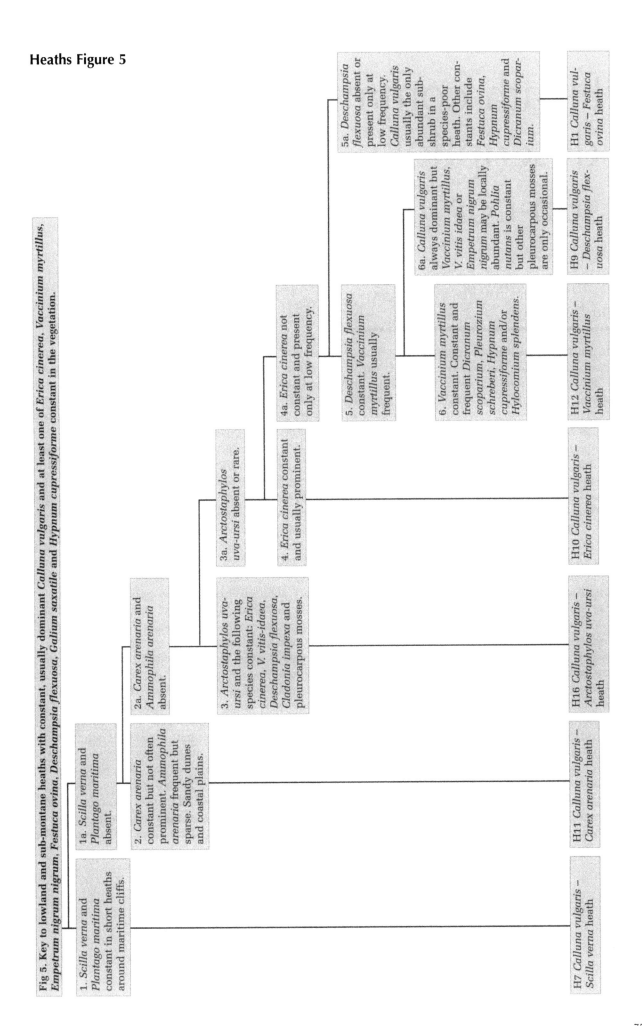

Fig 5. Key to lowland and sub-montane heaths with constant, usually dominant *Calluna vulgaris* and at least one of *Erica cinerea*, *Vaccinium myrtillus*, *Empetrum nigrum*, *Festuca ovina*, *Deschampsia flexuosa*, *Galium saxatile* and *Hypnum cupressiforme* constant in the vegetation.

1. *Scilla verna* and *Plantago maritima* constant in short heaths around maritime cliffs.

1a. *Scilla verna* and *Plantago maritima* absent.

2. *Carex arenaria* constant but not often prominent. *Ammophila arenaria* frequent but sparse. Sandy dunes and coastal plains.

2a. *Carex arenaria* and *Ammophila arenaria* absent.

3. *Arctostaphylos uva-ursi* and the following species constant: *Erica cinerea*, *V. vitis-idaea*, *Deschampsia flexuosa*, *Cladonia impexa* and pleurocarpous mosses.

3a. *Arctostaphylos uva-ursi* absent or rare.

4. *Erica cinerea* constant and usually prominent.

4a. *Erica cinerea* not constant and present only at low frequency.

5. *Deschampsia flexuosa* constant. *Vaccinium myrtillus* usually frequent.

5a. *Deschampsia flexuosa* absent or present only at low frequency. *Calluna vulgaris* usually the only abundant sub-shrub in a species-poor heath. Other constants include *Festuca ovina*, *Hypnum cupressiforme* and *Dicranum scoparium*.

6. *Vaccinium myrtillus* constant. Constant and frequent *Dicranum scoparium*, *Pleurozium schreberi*, *Hypnum cupressiforme* and/or *Hylocomium splendens*.

6a. *Calluna vulgaris* always dominant but *Vaccinium myrtillus*, *V. vitis idaea* or *Empetrum nigrum* may be locally abundant. *Pohlia nutans* is constant but other pleurocarpous mosses are only occasional.

H7 *Calluna vulgaris – Scilla verna* heath

H11 *Calluna vulgaris – Carex arenaria* heath

H16 *Calluna vulgaris – Arctostaphylos uva-ursi* heath

H10 *Calluna vulgaris – Erica cinerea* heath

H12 *Calluna vulgaris – Vaccinium myrtillus* heath

H9 *Calluna vulgaris – Deschampsia flexuosa* heath

H1 *Calluna vulgaris – Festuca ovina* heath

5 Heath community descriptions and sub-community keys

H1 *Calluna vulgaris – Festuca ovina* heath

This is a heather-dominated community which is very poor in vascular associates, although sometimes showing a modest diversity among the bryophytes and, more especially, the lichens. *Calluna vulgaris* is usually the only woody species and the most abundant. The height and cover of the canopy are very variable depending on the age of the heather and the consequent phase of development, and also on grazing intensity. *Erica cinerea*, *Ulex minor* and *U. gallii*, important in dry heaths further south and west, are largely excluded as is *Erica tetralix*. *Ulex europaeus* is uncommon, except where there has been disturbance.

Typically there are no grassy areas but *Festuca ovina* is very common throughout, usually as scattered tussocks, often with less than 30% total cover. Other grasses are few. *Agrostis capillaris* is occasional and there may be a little *Deschampsia flexuosa*. Associated dicotyledons are also few and patchy. Occasionally *Senecio jacobaea*, *Galium saxatile*, *Cerastium fontanum*, *Campanula rotundifolia* and *Luzula campestris* may be present. Two species locally important in particular situations are *Pteridium aquilinum* and *Carex arenaria*. In bare areas *Rumex acetosella* together with ephemerals such as *Aphanes arvensis*, *Teesdalia nudicaulis*, *Myosotis ramosissima* and the annual *Aira praecox* may be found.

Only a few bryophyte species occur throughout the community. *Hypnum cupressiforme* and *Dicranum scoparium* are both constant. These two usually form the bulk of the bryophyte cover in both pioneer and degenerate *Calluna*. In such situations *Hylocomium splendens*, *Pleurozium schreberi*, *Ptilidium ciliare* and *Dicranella heteromalla* are occasional. Lichens may exceed mosses in cover. *Cladonia* species are prominent with encrusting species such as *Cladonia pyxidata*, *C. squamosa* and *C. fimbriata* on bare ground. Species like *C. impexa*, *C. furcata* and *C. arbuscula* are especially abundant on old *Calluna* together with *Hypogymnia physodes*.

This community is confined to acid, base-poor and oligotrophic sandy soils in the more continental lowlands of eastern England. The profiles under the community are usually brown sands which are free to excessive-draining and have a low surface pH. In some localities, such as in Lincolnshire and around the Weald, the impoverished soils are derived from arenaceous bedrock, but they have mostly developed from sandy glacio-fluvial drift, sometimes supplemented by aeolian sand. The community occurs through the eastern lowlands of England, although it is now very local.

This heath has been traditionally managed with burning and grazing (both domestic livestock and wild herbivores such as rabbits and deer). However in many areas the abandonment of this traditional management has been followed by agricultural improvement or afforestation which has reduced and fragmented tracts of this community. In other areas the lack of grazing and burning has often permitted seral progression to scrub and woodland. The most common woody invaders are *Betula pendula* and *Pinus* spp., and more occasionally, *Quercus robur* if mature trees are fairly close by.

H1

Lichens poorly represented.

Hypnum cupressiforme very common among collapsed *Calluna vulgaris* and in grassy patches.

Cover of *Calluna vulgaris* rather open with some of *Cladonia uncialis, C. fimbriata, C. pyxidata, C. impexa, C. squamosa, Cornicularia aculeata* and *Hypogymnia physodes* locally abundant on mor and bare ground.

Bryophytes rather patchy but lichens often extensive with *Cladonia furcata* and *C. macilenta* common.

Teucrium scorodonia frequent with *Senecio jacobaea, Agrostis capillaris* and *Galium saxatile* occasional.

Carex arenaria constant, sometimes abundant.

Very impoverished rank canopies of *Calluna vulgaris.*

H1a

Hypnum cupressiforme sub-community

Cover of *Calluna vulgaris* is often less than complete and only moderately tall. The most prominent associates are *Festuca ovina* and the cryptogams which often exceed 50% cover. Among the mosses *Hypnum cupressiforme* and *Dicranum scoparium* are very frequent. Lichens are also abundant, *Cladonia pyxidata, C. squamosa, C. fimbriata* and *C. gracilis* usually predominate, bulkier species like *C. impexa, C. uncialis* and *C. arbuscula* being less frequent and *Hypogymnia physodes* occasional.

H1b

Hypogymnia physodes – Cladonia impexa sub-community

The heather cover is the same as in H1a but degenerate bushes predominate. The contribution of *Festuca ovina, Hypnum cupressiforme* and *Dicranum scoparium* is more uneven, the mosses occurring as small patches among collapsed stems. Lichens by contrast are abundant in these areas, encrusting species being joined or exceeded by *Cladonia impexa, C. furcata* and *C. macilenta. Hypogymnia physodes* is most frequent here.

H1c

Teucrium scorodonia sub-community

Degenerate bushes are scarce and heather cover is generally vigorous and extensive. Within more open areas scattered plants of *Teucrium scorodonia* or *Senecio jacobaea* can be found or small stretches of turf with *Festuca ovina, Agrostis capillaris, Deschampsia flexuosa* and scattered *Rumex acetosella, Galium saxatile* and *Cerastium fontanum*. Lichens are very infrequent, but *Hypnum cupressiforme* and *Dicranum scoparium* remain common.

H1d

Carex arenaria sub-community

The heather is often tall and somewhat open with scattered plants or denser patches of *Carex arenaria,* sometimes co-dominant and growing on sandy soil profiles. *Festuca ovina* occurs sparsely sometimes with a little *Rumex acetosella,* but other herbs are scarce. Among the cryptogams only *Hypnum cupressiforme* is frequent. This sub-community is found on coastal and inland dune systems.

H1e

Species-poor sub-community

Unbroken canopies of dense and often tall heather, up to 50 cm or more high, are characteristic here, among which virtually no associates can survive, apart from occasional sparse plants of *Hypnum cupressiforme.* This sub-community is typically of building and mature heather.

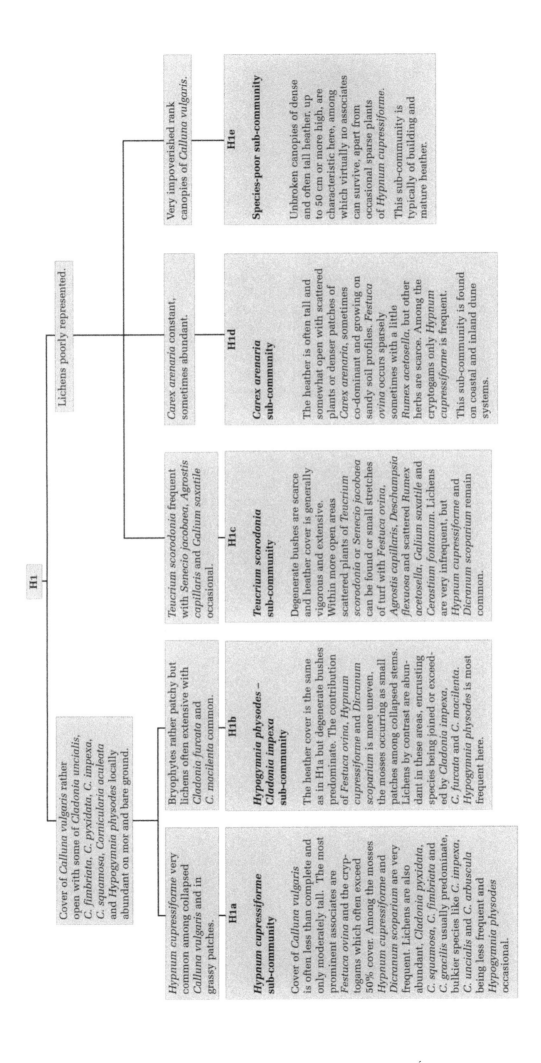

H2 *Calluna vulgaris – Ulex minor* heath

This community is generally dominated by *Calluna vulgaris*, but with both *Erica cinerea* and *Ulex minor* playing a very frequent and sometimes prominent role in the sub-shrub layer. The constancy of the latter two species provides the most obvious floristic distinction between this community and the *Calluna vulgaris – Festuca ovina* heath (H1). The canopy is very variable in height, from 10 cm to 80 cm or more, and the structure depends greatly on the growth phase of the *Calluna* and whether the individuals are of even or uneven age. Where burning occurs (for example, in the New Forest) a characteristic patchwork of swales is formed. After fire *E. cinerea* often increases in frequency because of its prolific seeding. *Ulex minor* normally plays a subsidiary role, forming a patchy understorey below the *Calluna*. No other sub-shrubs are found throughout the community. *Ulex europaeus* is occasional, but may be locally abundant after disturbance. *Erica tetralix* and *Vaccinium myrtillus* are found in particular sub-communities.

Other consistent associates are very few. *Deschampsia flexuosa* is constant but patchy, though it may be more common in grazed areas. On wetter soils it tends to be replaced by *Molinia caerulea*. *Pteridium aquilinum* is occasional overall and preferential for one sub-community. Scattered plants of *Potentilla erecta* or *Galium saxatile* may also be found in more open areas. In some stands invading seedling and sapling trees may occur, particularly *Quercus* spp., *Betula* spp. and *Pinus sylvestris*.

In the absence of burning, mosses and lichens can become common. *Dicranum scoparium* and *Hypnum jutlandicum* are the most frequent bryophytes, with peat-encrusting *Cladonia* species and larger species such as *Cladonia furcata* and *C. arbuscula*. *Hypogymnia physodes* can sometimes be found on old heather stems.

This community is characteristic of impoverished acid soils, predominantly free-draining in south-east and central southern England. It is characteristic of free-draining profiles developed from pervious arenaceous or pebbly parent materials. Typically, these parent materials have given rise to some kind of podzolic profile under this community which is highly acidic, often with a superficial pH between 3.5 and 4.5, and generally impoverished. The community occurs from the Poole Harbour area in the west through the New Forest, where stands are particularly numerous and extensive, to Surrey and the High Weald in the east, where it occurs as more local and fragmented tracts of heathland.

The vegetation takes much of its structural, and some of its floristic, character from traditional grazing and burning treatments. However, when released from these treatments a progression to woodland can be expected. Agricultural improvements and forestry have caused fragmentation and isolation of small remnants of this community in many places.

H2

Vaccinium myrtillus constant and locally abundant, Pteridium aquilinum frequent and with scattered tree seedlings and saplings.

Pteridium aquilinum can be quite common, but Vaccinium myrtillus and young trees very scarce.

Molinia caerulea occasional at most with Deschampsia flexuosa very common. Dicranum scoparium and Hypnum ericetorum with Cladonia spp. patchily prominent in more open areas.

Molinia caerulea very common, often almost totally displacing Deschampsia flexuosa. Erica tetralix frequent and sometimes exceeding E. cinerea.

H2b

Vaccinium myrtillus sub-community

The sub-shrub canopy usually consists of mixtures of heather with sometimes substantial amounts of Ulex minor, Erica cinerea and, strongly preferential, Vaccinium myrtillus. Pteridium aquilinum is more frequent, and in open areas there is usually some Deschampsia flexuosa and occasionally a little Molinia caerulea. Young trees are also strongly preferential, with oak and birch seedlings and saplings frequent, and small pines locally prominent. Hypnum jutlandicum and Dicranum scoparium are occasional but lichens are sparse.

This sub-community tends to be found at the higher altitudes in the range of H2.

H2c

Molinia caerulea sub-community

Calluna vulgaris is abundant, usually with smaller amounts of Ulex minor and particularly of Erica cinerea which may be joined with or replaced by Erica tetralix. More obviously preferential is Molinia caerulea. Pteridium aquilinum is uncommon and young trees are rarely found. The ground layer is also sparse.

This sub community is predominantly found on soils with impeded drainage.

H2a

Typical sub-community

Calluna vulgaris is generally strongly dominant with subsidiary amounts of Ulex minor and Erica cinerea, the latter very variable and sometimes absent. Neither Vaccinium myrtillus nor Erica tetralix occur and Molinia caerulea is scarce. Deschampsia flexuosa is common and sometimes with or replaced by Festuca rubra. Pteridium aquilinum and tree seedlings are infrequent. Bryophytes and lichens can be conspicuous among older heather or after burning, with Cladonia fimbriata, C. coccifera, C. chlorophaea and C. arbuscula all slightly preferential.

H3 *Ulex minor – Agrostis curtisii* heath

This community contains nearly all the sub-shrub vegetation in which *Ulex minor* and *Agrostis curtisii* occur together as important components, although *A. curtisii* can also occur occasionally in *Calluna vulgaris – Ulex minor* heath (H2). The canopy is usually fairly low, 20-30 cm high, and *Calluna* frequently dominates, especially when it has not been burned for some time. Compared with less oceanic heaths the most unusual feature of the woody cover is the occurrence together of *Erica cinerea* and *E. tetralix*, both of which are constants. Both can grow vigorously, although *E. cinerea* is likely to be more prominent especially after burning. *Erica tetralix* can have high cover locally, especially on more strongly gleyed soils. *Ulex minor* maintains its high frequency throughout, although its abundance is very variable. In stands which have not been burnt for some time *A. curtisii* and *Molinia caerulea*, the two characteristic and constant grasses, are generally scattered, but after burning *A. curtisii* and to a lesser degree *Molinia* can become prominent. In contrast to H2 *Deschampsia flexuosa* is very scarce. *Pteridium aquilinum* occurs occasionally and other herbs are found as scattered individuals. *Potentilla erecta*, *Polygala serpyllifolia*, *Carex pilulifera* and the parasitic *Cuscuta epithymum* can all be found occasionally. In disturbed or burned situations *Viola lactea* is very characteristic.

Burning has the effect of opening the canopy, and mosses and lichens become prominent. Among the bryophytes *Campylopus brevipilus* is most distinctive and can be accompanied by *C. paradoxus*, *Polytrichum juniperinum*, *Dicranum scoparium*, *Hypnum jutlandicum* and *Leucobryum glaucum*. *Cladonia impexa* is one of the most common and conspicuous lichens with peat-encrusting species such as *C. floerkeana*, *C. coccifera* and *C. pyxidata*. *Hypogymnia physodes* often colonises old heather stems.

This community is the characteristic sub-shrub community of impoverished acid soils which are protected against parching by a measure of drainage impedance and a moderately oceanic climate. It occupies a distinct position on soils that are too dry for the *Erica tetralix – Sphagnum compactum* wet heath (M16) and too moist for the *Calluna vulgaris – Ulex minor* heath (H2). It is largely confined to south Dorset and Hampshire.

The combination of drainage impedance and climate is the major influence on the floristics of this community, although grazing and burning still often exert an important measure of control on its composition and structure. The general effect of the combination of these treatments is to curtail the mature and degenerate phase of *Calluna* and to set back repeatedly any invasion of trees and seral progression to woodland. The abandonment of traditional land use and soil improvement for agriculture in many stands of this type of vegetation has meant that surviving tracts can be fragmented, and are often sharply delineated from their surrounds.

H3

Calluna vulgaris and *Erica tetralix* reduced in frequency and cover and *Molinia caerulea* and *Ulex minor* somewhat patchy. *Agrostis curtisii* very abundant with *Ulex europaeus* common and *Viola lactea* often persistent.

Agrostis curtisii very frequent, but not extensive, and *Ulex europaeus* occasional at most among mixed or *Calluna vulgaris*-dominated canopies.

Species opposite very sparse among usually dense sub-shrub canopies.

H3a

Typical sub-community

The sub-shrubs typically form an extensive canopy, often with *Calluna vulgaris* as the main dominant, although, sometimes more mixed. Grasses are usually subordinate. *Ulex europaeus* is occasional and *Erica ciliaris* is found in south Dorset vegetation. Other species are few, although *Potentilla erecta* is preferential at low frequency, and there is sometimes a little *Pteridium aquilinum*, *Carex pilulifera*, *Polygala serpyllifolia* or *Cuscuta epithymum*. Bryophytes and lichens are sparse.

Polygala serpyllifolia common with a patchy cover of bryophytes and lichens on more open areas. Species present include *Caampylopus brevipilus*, *Polytrichum juniperinum*, *Cladonia impexa*, *C. floerkeana* and *C. coccifera*.

H3b

Cladonia spp. sub-community

The sub-shrub canopy is somewhat open, and although *Calluna vulgaris* is the leading species, dominance is often shared between woody plants and *Polygala serpyllifolia*, a strong preferential here. There is a noticeable cover of mosses and lichens, the species listed above being most frequent.

H3c

Agrostis curtisii sub-community

Agrostis curtisii is very abundant and *Erica cinerea* often co-dominant, but *Ulex minor* and *Molinia caerulea* are patchy and *Calluna vulgaris* and *Erica tetralix* much reduced in frequency and cover. *Ulex europaeus* is strongly preferential in this sub-community where disturbance, often by burning, is characteristic. The early stages of development often allow colonisation by *Viola lactea*, which then persists as the dominants expand.

H4 *Ulex gallii – Agrostis curtisii* heath

This community is very similar to *Ulex minor – Agrostis curtisii* heath (H3), with the replacement of one gorse by another. The western limit of *U. minor* in east Dorset forms the boundary between these two heath types. Apart from this difference they share five constants, namely, *Calluna vulgaris*, *Erica cinerea*, *E. tetralix*, *Molinia caerulea* and *Agrostis curtisii*, and these species, together with *U. gallii*, generally account for the bulk of the vascular cover. Their proportions and structure, however, vary considerably so that the appearance of stands differs markedly. The vegetation can vary from a short mixed canopy of grasses and sub-shrubs no more than 10 cm high (grass heath), to a canopy of woody plants 50 cm or more high. There may also be quite extensive areas of barer ground. *Calluna* and *U. gallii* are the most common species and are often abundant. *Calluna* often dominates. The frequent occurrence of *E. cinerea* and *E. tetralix* together distinguishes this vegetation from the corresponding dry heath *Calluna vulgaris – Ulex gallii* heath (H8). Four other sub-shrubs are more restricted. *Vaccinium myrtillus* is commoner at higher altitudes with increased rainfall. The others, *Salix repens*, *Erica ciliaris* and *E. vagans* (a species restricted to the Lizard in Cornwall) are found in the wetter *E. tetralix* sub-community.

Two grasses are constant, *Agrostis curtisii* and *Molinia caerulea*, which always make some contribution to the cover. Among the grassier heaths, *Festuca ovina* and *Danthonia decumbens* are particularly important, with the sedges *Carex binervis* and *C. pilulifera* also characteristic. On cooler, moister slopes *Scirpus cespitosus* can be prominent. The only dicotyledonous herb which is a constant of this community is *Potentilla erecta*, which occurs as scattered individuals. Other occasional herbs are *Polygala serpyllifolia* and *Pedicularis sylvatica*, with *Viola lactea* in disturbed situations. There are a variety of bryophytes and lichens, but none occur with any frequency.

This community is confined to the warm oceanic parts of south-west Britain where it occurs on a variety of moist, acid soils. Like its eastern counterpart, H3, this is a vegetation type of acid soils that are too moist for dry heath but not so consistently waterlogged as to be able to sustain wet heath. The community is confined to south-west Britain, beyond a line from mid-Dorset to the Quantocks, and including parts of the south Wales seaboard up to altitudes of 500 m.

Both climatic and edaphic conditions combine to influence the general character of this vegetation. However, in most situations burning and grazing have a marked effect on the floristics and physiognomy of the vegetation and, with the exception of situations such as the Lizard where exposure to high and frequent winds is combined with a scarcity of seed parents, these treatments are important for maintaining the community against succession to woodland. As with many lowland heath communities intensive improvement for agriculture and afforestation has reduced and fragmented its extent.

H4

Erica tetralix very infrequent in vegetation usually dominated by *Agrostis curtisii* or *E. cinerea.*

Erica tetralix a frequent component of the sub-shrub community.

Erica cinerea rather patchy but *Vaccinium myrtillus* common in a grassy heath with *Festuca ovina, Danthonia decumbens, Agrostis capillaris, Galium saxatile* and occasional *Carex pilulifera* and *C. binervis.*

Erica cinerea and *E. tetralix* both very common, but associates listed opposite occasional at most in a less grassy heath.

Scirpus cespitosus constant in vegetation usually dominated by *Calluna vulgaris,* but with other sub-shrubs and grasses locally abundant.

Scirpus cespitosus absent from vegetation variously dominated by one or more sub-shrubs and grasses and with *Erica vagans* or *E. ciliaris* locally abundant.

H4a

Agrostis curtisii – Erica cinerea sub-community

Unusually there is an almost total absence of *Erica tetralix.* Otherwise all the constants remain frequent except for *Agrostis curtisii* which is often dominant, forming a virtually pure and tussocky sward. Other grasses do not generally increase in cover. Other vascular and cryptogam associates are few in number.

This sub-community occurs throughout the range of H4 and is often present as regenerating vegetation after burning.

H4b

Festuca ovina sub-community

Agrostis curtisii is often the most abundant species, most commonly with a fairly rich mixture of sub-shrubs and herbs forming a grass heath. Both *Ulex gallii* and *Calluna vulgaris* have high frequencies; *Erica cinerea* is more patchy and *Erica tetralix-* common but not constant. *Vaccinium myrtillus* is most common in this sub-community, often as sparse shoots. The associates listed above are more abundant and preferential. *Potentilla erecta* is very common, often with *Galium saxatile.* Bryophytes and lichens are very sparse in ungrazed stands but occasional in grazed vegetation.

This sub-community, along with the *Scirpus* sub-community, can be found at higher altitudes on Dartmoor and Exmoor and also throughout the range on free-draining soils with grazing.

H4d

Scirpus cespitosus sub-community

Scirpus cespitosus is constant and strongly preferential with very frequent *Calluna vulgaris, Molinia caerulea* and *Erica tetralix* and common *E. cinerea* and *Vaccinium myrtillus. Ulex gallii* and *Agrostis curtisii,* both constants, distinguish this sub-community from wet heath. *Dicranum scoparium* and *Leucobryum glaucum* are frequent and *Cladonia impexa* and *C. uncialis* occasional.

This sub-community, along with the *Festuca* sub-community, can be found at higher altitudes on Dartmoor and Exmoor.

H4c

Erica tetralix sub-community

Both grasses and sub-shrubs have very high frequency, including *Erica tetralix,* and each can be abundant. *Potentilla erecta* is frequent and *Danthonia decumbens, Polygala serpyllifolia* and *Carex panicea* occasional. This vegetation is best known on the Lizard where *Erica ciliaris* or *E. vagans* occur in this type of heath. Bryophytes and lichens show a varying representation.

This sub-community is especially characteristic of lower altitudes and is well represented on the Devon Pebble-Bed commons, the lower fringes of Dartmoor and Bodmin Moor and on the Lizard.

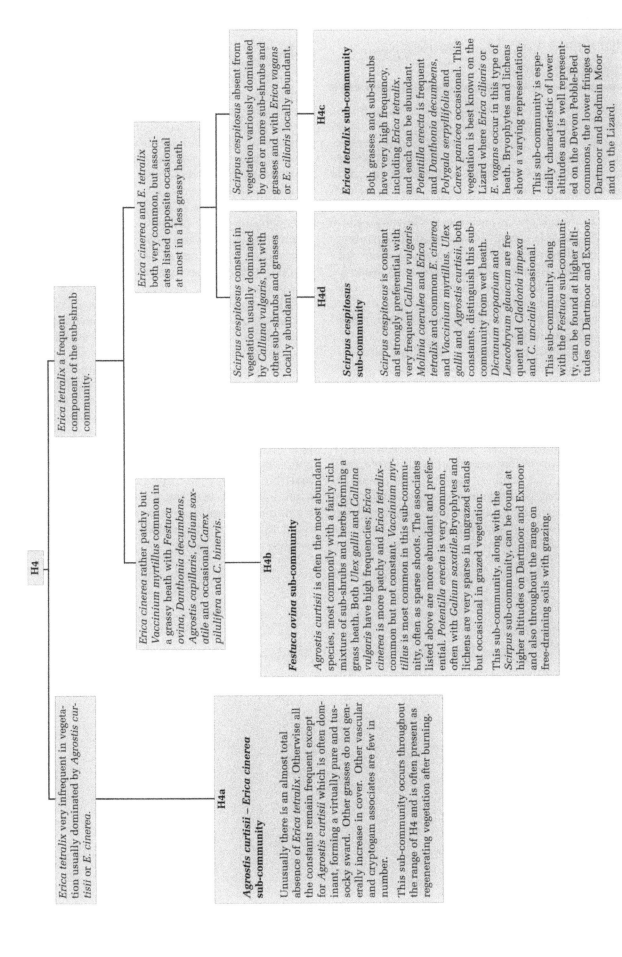

H5 *Erica vagans – Schoenus nigricans* heath

This heath is one of two sub-shrub communities in which the nationally-rare *Erica vagans* makes a constant and prominent contribution. *Schoenus nigricans* is also constant and usually abundant as strongly-developed tussocks. *Molinia caerulea* and *Erica tetralix* are also constant, often with high cover, and together these four species dominate in mixtures. Between these species there is a well-defined system of runnels giving a distinct microhabitat. Among other sub-shrubs only *Ulex gallii* occurs with any frequency and may be co-dominant. *Calluna* is only occasional and *E. cinerea* scarce. *Genista anglica*, however, can occur frequently and is preferential to this community.

In undisturbed stands which have not been burnt or grazed recently, *Schoenus* and *E. vagans* tend to be dominant and the vegetation is choked with their litter. Then, even common associates like *Potentilla erecta* and *Festuca ovina* can be crowded out. After burning or grazing, or both, the associated flora is much richer. In the wetter runnels sedges are often important with *Carex pulicaris* constant, *C. panicea* and *C. flacca* frequent and, on gabbro, *C. hostiana*. *Anagallis tenella*, also constant, may form extensive mats. Among taller herbs *Serratula tinctoria* and *Succisa pratensis* are constant and *Sanguisorba officinalis* frequent.

There are a number of occasional associates. On wetter ground *Phragmites australis* can be present as conspicuous but scattered shoots.

Bryophytes vary considerably among stands but *Campylium stellatum* is constant and very frequent in runnels and may be abundant, often with *Riccardia multifida*, *R. sinuata* and, over gabbro, *Scorpidium scorpioides*. After wet weather runnels often have swollen gelatinous globules of blue-green algae.

This community is confined to wet, base-rich but calcium-poor mineral soils and shallow peats on the Lizard in Cornwall. Here the distinctive parent materials of serpentine and gabbro found in this area have given rise to soils that have a superficial pH of between 5.5 and 7.5 but in which magnesium predominates over calcium. The community makes the major proportion of the open and enclosed heaths of the hinterland of the peninsula.

The floristics of this community are influenced both by the mild oceanic climate and underlying bedrocks of serpentine and gabbro, but the composition and physiognomy of particular stands are affected by burning and sometimes also by grazing. Other past treatments like the cutting of turf have also probably influenced the appearance and distribution of this community. There have been losses of this vegetation type to modern techniques of land improvement and much of the remaining extent has statutory or voluntary protection.

H5

Frequent *Eleocharis multicaulis*, *Eriophorum angustifolium*, *Drosera rotundifolia*, *Pinguicula lusitanica*, and *Dactylorhiza incarnata incarnata* growing in runnels that are usually flooded for much of the year. In ungrazed stands, *Phragmites australis* may be locally abundant.

Vegetation variable in composition and structure, but species listed opposite rare.

H5b

***Eleocharis multicaulis* sub-community**

Schoenus nigricans dominant with *Molinia caerulea* and *Erica vagans*. *Erica tetralix* is somewhat less abundant. *Calluna vulgaris* and *Erica tetralix* usually absent and *Carex panicea* is also typically missing.

H5a

Typical sub-community

This vegetation has all the general features of the community with no additional preferential species. The tussock/runnel structure is often well-defined, but species-richness depends greatly on treatment and especially time since burning.

H6 *Erica vagans – Ulex europaeus* heath

This community is a distinctive type of sub-shrub vegetation, but rather variable in floristics and structure. The most obvious feature is a mixed canopy of sub-shrubs in which *Erica vagans* and *Ulex europaeus* are the usual co-dominants. The canopy is generally 30-60 cm high but in exposed situations may be not more than 10 cm high. Two other constant sub-shrubs, *Ulex gallii* and *E. cinerea*, can also be abundant although the former may be suppressed in dense stands. *Calluna vulgaris* is not frequent and has generally low cover.

In contrast to the *Erica vagans – Schoenus nigricans* heath (H5) community, *E. tetralix* is only occasional and confined to wetter soils (see sub-community H6d) with several preferential associates. The only herbaceous associates common throughout are *Carex flacca*, *Potentilla erecta* and *Polygala vulgaris*. The most common and distinctive herbs of this community are *Viola riviniana*, *Filipendula vulgaris*, *Stachys betonica*, *Hypochoeris radicata*, *Agrostis canina* ssp. *montana*, *Dactylis glomerata* and *Scilla verna*. Most of these species are found in recently burned stands but become more scattered and reduced in number as the vegetation and litter increase.

On shallower soils, especially when grazed, a rich short herb layer is maintained with several additional species including *Festuca ovina*, *Thymus praecox*, *Lotus corniculatus*, *Galium verum*, *Jasione montana*, *Danthonia decumbens* and *Brachypodium sylvaticum*. Immediately after burning, diversity is increased, with ephemerals including *Aira caryophyllea* and *Centaurium erythraea*. Continued burning and the dense shade and litter of older stands inhibit bryophytes and lichens, which as a result are uncommon.

This community is confined to the Lizard in Cornwall where it is characteristic of free-draining brown earths that are usually quite base-rich but calcium-poor and fairly oligotrophic. It is found on soils similar to that of H5 with a pH of generally between 5 and 7, but which are more free-draining. Therefore it is typically found on the steeper, shedding slopes around coves and on the cliff tops of the headlands. Although it is mainly coastal in distribution it is not strictly speaking a maritime heath and is replaced on slopes which are exposed to salt spray by *Calluna vulgaris – Scilla verna* heath (H7).

Edaphic variation and local differences in the warm oceanic climate strongly influence floristic diversity, but treatments, especially burning, and to a lesser extent grazing, also have a marked effect on composition and physiognomy of the vegetation. However, the progression to scrub and woodland in the absence of these treatments would probably be slow due to the lack of seed parents and the poor quality of the soil. Preferential cultivation of the more fertile soils developed over gabbro and schists means that the community survives most extensively over serpentine.

H6

Molinia caerulea and Potentilla erecta constant with Serratula tinctoria frequent, but Filipendula vulgaris rather uncommon.

Filipendula vulgaris very frequent but Molinia caerulea absent and Serratula tinctoria erecta and Serratula tinctoria only occasional.

Agrostis curtisii constant and often abundant, particularly after burning, with Calluna vulgaris scarce. Carex panicea, Hypericum pulchrum, Viola lactea and Polygala serpyllifolia frequent.

Agrostis curtisii and associates listed opposite absent, but Erica tetralix constant in small amounts and Sanguisorba officinalis and Schoenus nigricans frequent.

Sub-shrub canopy usually low and open with a rich flora between the bushes including Danthonia decumbens, Koeleria macrantha, Aira caryophyllea, Galium verum and Leontodon taraxacoides as constants.

Sub-shrub canopy usually extensive with occasional Rubus fruticosus agg., Prunus spinosa and Pteridium aquilinum, and scattered Teucrium scorodonia and Geranium sanguineum, but associates listed opposite very sparse in more recently regenerating stands.

H6c

Agrostis curtisii sub-community

Apart from occasional occurrences in H6a, Agrostis curtisii is largely confined to this sub-community where it can be very abundant especially after burning. Molinia caerulea is also frequent and the two grasses sometimes dominate under an open canopy of Erica vagans, E. cinerea, Ulex gallii and U. europaeus. Among the herbs Viola riviniana is absent and Filipendula vulgaris scarce. Danthonia decumbens and Potentilla erecta are very frequent and Stachys betonica and Serratula tinctoria are common, as are the preferential species listed above.

H6d

Molinia caerulea sub-community

Erica vagans and Ulex europaeus retain high frequency and abundance with smaller amounts of Ulex gallii. Both Erica cinerea and Calluna vulgaris are reduced in frequency but Erica tetralix is constant in small amounts. In stands not recently burned Molinia caerulea is distinctive. Its litter depresses herbs and generally only Carex flacca, Viola riviniana, Potentilla erecta and Stachys betonica occur with any frequency, with small amounts of the preferentials listed above.

H6b

Festuca ovina sub-community

Here the abundance and height of the sub-shrub canopy is less than in the typical form with a total cover of often less than 50%. Apart from Potentilla erecta the common herbs are all well represented in this sub-community. In the more open conditions there are more cryptogams than usual, Hypnum cupressiforme s.l. is frequent and various Cladonia spp. common.

H6a

Typical sub-community

Typical heath with a complete range of floristic and structural vegetation related to burning. A few years into the cycle there is a well-developed sub-shrub canopy with most of the community constants. Richer stands have occasional Teucrium scorodonia and Geranium sanguineum, the only preferential dicotyledons here. With increasing age the constant herbs begin to thin out leaving a dense woody cover.

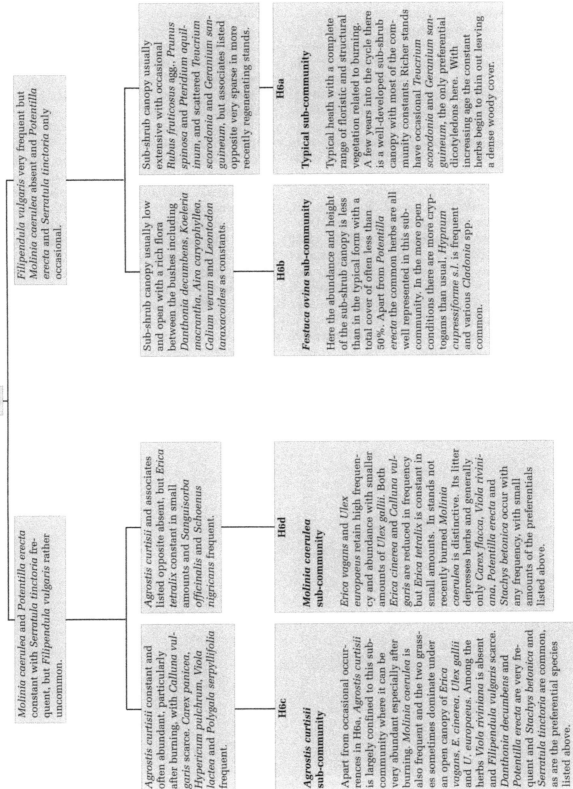

H7 *Calluna vulgaris – Scilla verna* heath

In this heath, sub-shrubs are a consistent feature, though they are not always obvious. The canopy is typically very short, rarely over 20 cm, and sometimes when grazed forming a mat only 2-3 cm high. The cover of woody plants is rarely continuous. Even where sub-shrubs are more extensive, they are commonly penetrated by herbs. *Calluna vulgaris* is the most frequent sub-shrub and the commonest dominant, though on dry soils it is accompanied by *Erica cinerea*. On wetter soils the latter is much reduced and *E. tetralix* and/or *Empetrum nigrum* ssp. *nigrum* are the usual associates. No other woody species occurs frequently throughout, although *Ulex gallii* is occasional.

Among herbaceous associates grasses are often important. *Festuca ovina* is the most frequent grass species, though *F. rubra* is also common. Also common and a constant is *Holcus lanatus*, often with *Dactylis glomerata* on drier soils or *Danthonia decumbens* on moister ground. In wetter, northern heaths *Agrostis capillaris* and *Anthoxanthum odoratum* can become very common, but *Molinia caerulea* is infrequent. There are a variety of other herbs. Most distinctive among the constants are *Plantago maritima* and *Scilla verna*. Other common and constant species are *Plantago lanceolata*, *Potentilla erecta*, *Lotus corniculatus*, *Thymus praecox* and *Hypochoeris radicata*, the latter of which tends to favour drier soils. *Anthyllis vulneraria* also favours drier soils as do *Euphrasia* species. Other species are more characteristic of particular sub-communities.

In contrast cryptogams are few and never show high cover. Among the mosses only *Hypnum cupressiforme s.l.* is moderately frequent and *Frullania tamarisci*, the commonest hepatic, is infrequent. Several *Cladonia* species are occasional.

This community occurs over a wide variety of moderately base-poor soils on the less exposed parts of maritime cliffs all around the coast of Britain except to the east and south between Durham and Dorset. The single most distinctive difference between the habitat of this kind of heath and the habitats of other sub-shrub communities is the input of salt spray generated by breaking waves and carried inland by the wind.

The floristic and structural variation in this community is influenced by the climatic and edaphic differences both throughout the considerable geographic range of the community and over particular stretches of cliff. Grazing also affects the composition and appearance of the vegetation and probably contributes to maintaining it against successional change. However, over much of its range this vegetation can be considered a climatic climax as exposure to even small amounts of salt spray hinders the invasion of woody invaders.

H7

Erica cinerea and Hypochoeris radicata reduced in frequency but Erica tetralix and Empetrum nigrum common. Plantago maritima and P. lanceolata often very conspicuous. Anthoxanthum odoratum and Agrostis capillaris frequent and Carex panicea and Carex nigra occasional.

Agrostis capillaris and Anthoxanthum odoratum occasional but Erica cinerea and Hypochoeris radicata remain very frequent and E. tetralix, Empetrum nigrum, Carex panicea and C. nigra are rare.

Erica tetralix constant with occasional Empetrum nigrum and frequent Danthonia decumbens and Succisa pratensis. Molinia caerulea, Nardus stricta and Salix repens locally prominent.

Empetrum nigrum constant, with Erica tetralix, Danthonia decumbens and Succisa pratensis only very occasional but Trifolium repens and Luzula multiflora quite common.

Armeria maritima and Sedum anglicum constant and often abundant with frequent Dactylis glomerata, Anthyllis vulneraria and Jasione montana, and occasional Plantago coronopus and Silene vulgaris maritima.

Dactylis glomerata and Anthyllis vulneraria occasional but other listed associates rare.

Viola riviniana, Polygala vulgaris, Carex flacca and C. caryophyllea frequent. Achillea millefolium, Leontodon taraxacoides, Galium verum, Stachys betonica and Serratula tinctoria occasional and Ulex europaeus locally prominent.

Viola riviniana occasional but other listed associates rare among an often impoverished Calluna vulgaris-dominated cover.

H7c
Erica tetralix sub-community

Two features distinguish this type of heath. First, among the sub-shrubs Erica cinerea is much reduced and replaced by Erica tetralix as the usual companion to Calluna vulgaris. The canopy is typically extensive, but short because of grazing. Secondly, although grasses are prominent Festuca ovina is often replaced by Festuca rubra as the most abundant species and Agrostis capillaris, Anthoxanthum odoratum and Danthonia decumbens are very common and sometimes co-dominant. Plantago maritima is often abundant with P. lanceolata common.

This sub-community is most prominent on the north-west coast of Britain, particularly in the Hebrides and Sutherland, with scattered occurrences down to Anglesey.

H7d
Empetrum nigrum ssp. nigrum sub-community

This type of heath shares several features with H7c; Erica cinerea is seldom found but is replaced here by Empetrum nigrum which is often co-dominant with Calluna vulgaris. Erica tetralix and Ulex species are rare. Festuca ovina, F. rubra, Agrostis capillaris and Anthoxanthum odoratum are frequent and Carex panicea common. Among dicotyledons Plantago maritima and P. lanceolata are very frequent and Thymus praecox and Lotus corniculatus common, while Hypochoeris radicata is very rare on the moist soils.

This sub-community is found predominantly in northern Britain.

H7a
Armeria maritima sub-community

Here, where salt-spray deposition is high, the canopy of sub-shrubs is generally less extensive, either with a mosaic of open areas or reduced to discrete patches of bushes. The only common sub-shrubs are Calluna vulgaris and Erica cinerea, although the latter is reduced in exposed situations. Growing among and between the bushes is Festuca ovina as the dominant, sometimes with Holcus lanatus and Dactylis glomerata. Apart from Potentilla erecta, which is scarce, all other community constants and the associates listed above are well represented.

This sub-community, along with the Viola sub-community, occurs throughout the range of H7 but is better developed to the south of Galloway with only local stations beyond this.

H7b
Viola riviniana sub-community

The sub-shrub canopy is more extensive than in H7a with both Calluna vulgaris and Erica cinerea very frequent and often co-dominant. Erica tetralix and Empetrum nigrum are typically very scarce. The herbaceous plants are most distinctive. All the community constants are well represented together with the species listed above.

This sub-community, along with the Armeria sub-community, occurs throughout the range of H7 but is better developed south of Galloway with only local stations north of this.

H7e
Calluna vulgaris sub-community

In general floristics this heath type resembles impoverished versions of the Armeria and Viola sub-communities though usually with a taller canopy. Calluna vulgaris is the usual dominant though Erica cinerea is common; all other sub-shrubs are scarce. Festuca ovina is the most frequent grass with F. rubra less common. Hypochoeris radicata is quite frequent but rosette herbs are poorly represented in the rank herbage. Thymus praecox and Lotus corniculatus, although common are often not abundant.

This sub-community is found throughout the range of H7 but is rarer to the north, although it is well-represented in Shetland.

H8 *Calluna vulgaris – Ulex gallii* heath

Floristically this is a diverse community with only three constants overall, namely *Calluna vulgaris*, *Erica cinerea* and *Ulex gallii*. *Erica tetralix*, *Molinia caerulea* and *Agrostis curtisii* are typically lacking from this community. Often the three constant sub-shrubs are co-dominant, but proportions are variable and where *E. cinerea* is reduced *Vaccinium myrtillus* can appear. On disturbed ground *U. europaeus* may be abundant and both *Pteridium aquilinum* and *Rubus fruticosus* agg. may appear in the heath.

Typically sub-shrub cover is high and herbs are sparse, but often the bushes are separated by grassy runnels, a feature accentuated by grazing. The most frequent grasses are *Agrostis capillaris* and *Festuca ovina* with *A. canina* ssp. *montana*, *F. rubra*, *Anthoxanthum odoratum* and *Danthonia decumbens* occasional to frequent. *Deschampsia flexuosa* and *Nardus stricta* are much more patchy in their occurrence. There is often some *Potentilla erecta* and *Galium saxatile*, and much more occasionally *Teucrium scorodonia* and *Polygala serpyllifolia*. Additional herbs are characteristic of particular sub-communities.

In general bryophytes and lichens are not numerous or diverse. There may be some *Hypnum cupressiforme* and *Dicranum scoparium*, and *Rhytidiadelphus squarrosus* and *Pleurozium schreberi* are more occasional. In more open situations, or on burned or disturbed bare ground, mosses such as *Campylopus paradoxus*, *Polytrichum piliferum* or *P. juniperinum* can become abundant along with lichens species such as *Cladonia impexa* and *C. squamosa.*

This community is found on free-draining, generally acid to circumneutral soils, in the warm oceanic regions of lowland Britain. It can be found over a wide range of arenaceous sedimentaries and acid igneous and metamorphic rocks as well as on silty and sandy superficials like loess and aeolian sands. The superficial pH underneath this community is usually from 3.5 to 4.5. It occurs throughout south-western England and Wales, on the Isle of Man and, more sporadically, in the southern Pennine fringes and the East Anglian coast.

Local climatic and edaphic conditions influence floristic variation; grazing by rabbits, sheep or cattle, and sometimes burning (which is normally an accidental occurrence), affect physiognomy and composition. The community is maintained against succession to woodland in most situations by grazing and burning, although in some situations exposure to the wind prevents the establishment of woody invaders such as *Betula* spp. and *Quercus* spp. Much former heath has been improved for agriculture and it now often survives as patches on marginal grazing land.

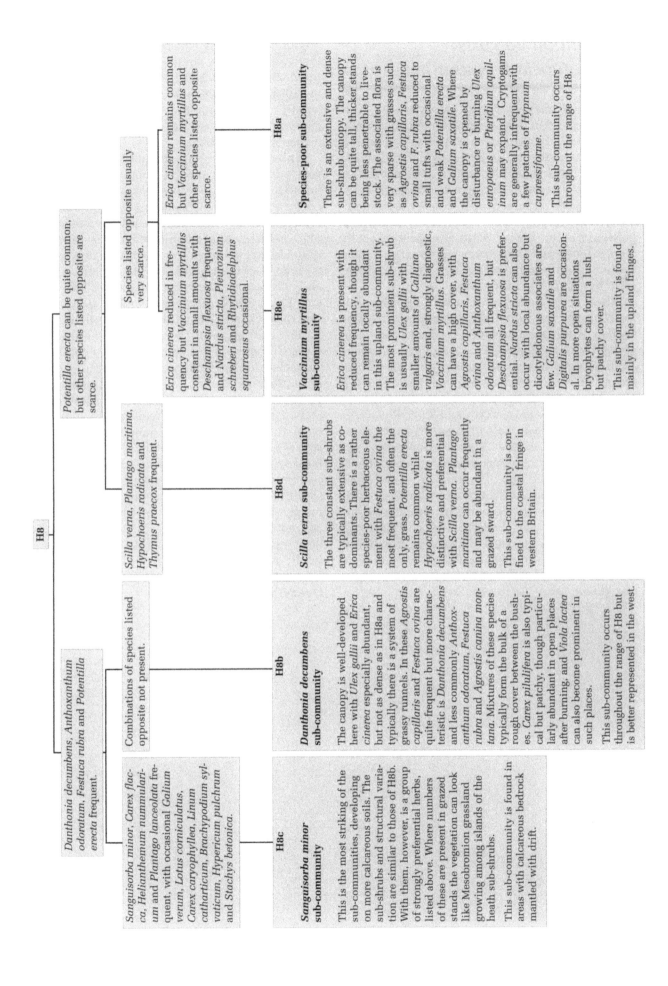

H8

Danthonia decumbens, Anthoxanthum odoratum, Festuca rubra and Potentilla erecta frequent.

Sanguisorba minor, Carex flacca, Helianthemum nummularium and Plantago lanceolata frequent, with occasional Galium verum, Lotus corniculatus, Carex caryophyllea, Linum catharticum, Brachypodium sylvaticum, Hypericum pulchrum and Stachys betonica.

Combinations of species listed opposite not present.

Potentilla erecta can be quite common, but other species listed opposite are scarce.

Scilla verna, Plantago maritima, Hypochoeris radicata and Thymus praecox frequent.

Species listed opposite usually very scarce.

Erica cinerea reduced in frequency but Vaccinium myrtillus constant in small amounts with Deschampsia flexuosa frequent and Nardus stricta, Pleurozium schreberi and Rhytidiadelphus squarrosus occasional.

Erica cinerea remains common but Vaccinium myrtillus and other species listed opposite scarce.

H8c

Sanguisorba minor sub-community

This is the most striking of the sub-communities, developing on more calcareous soils. The sub-shrubs and structural variation are similar to those of H8b. With them, however, is a group of strongly preferential herbs. listed above. Where numbers of these are present in grazed stands the vegetation can look like Mesobromion grassland growing among islands of the heath sub-shrubs.

This sub-community is found in areas with calcareous bedrock mantled with drift.

H8b

Danthonia decumbens sub-community

The canopy is well-developed here with Ulex gallii and Erica cinerea especially abundant, but not as dense as in H8a and typically there is a system of grassy runnels. In these Agrostis capillaris and Festuca ovina are quite frequent but more characteristic is Danthonia decumbens and less commonly Anthoxanthum odoratum, Festuca rubra and Agrostis canina montana. Mixtures of these species typically form the bulk of a rough cover between the bushes. Carex pilulifera is also typical but patchy, though particularly abundant in open places after burning, and Viola lactea can also become prominent in such places.

This sub-community occurs throughout the range of H8 but is better represented in the west.

H8d

Scilla verna sub-community

The three constant sub-shrubs are typically extensive as co-dominants. There is a rather species-poor herbaceous element with Festuca ovina the most frequent, and often the only, grass. Potentilla erecta remains common while Hypochoeris radicata is more distinctive and preferential with Scilla verna. Plantago maritima can occur frequently and may be abundant in a grazed sward.

This sub-community is confined to the coastal fringe in western Britain.

H8e

Vaccinium myrtillus sub-community

Erica cinerea is present with reduced frequency, though it can remain locally abundant in this upland sub-community. The most prominent sub-shrub is usually Ulex gallii with smaller amounts of Calluna vulgaris and, strongly diagnostic, Vaccinium myrtillus. Grasses can have a high cover, with Agrostis capillaris, Festuca ovina and Anthoxanthum odoratum all frequent, but Deschampsia flexuosa is preferential. Nardus stricta can also occur with local abundance but dicotyledonous associates are few. Galium saxatile and Digitalis purpurea are occasional. In more open situations bryophytes can form a lush but patchy cover.

This sub-community is found mainly in the upland fringes.

H8a

Species-poor sub-community

There is an extensive and dense sub-shrub canopy. The canopy can be quite tall. thicker stands being less penetrable to live-stock. The associated flora is very sparse with grasses such as Agrostis capillaris, Festuca ovina and F. rubra reduced to small tufts with occasional Potentilla erecta and weak Galium saxatile. Where the canopy is opened by disturbance or burning Ulex europaeus or Pteridium aquilinum may expand. Cryptogams are generally infrequent with a few patches of Hypnum cupressiforme.

This sub-community occurs throughout the range of H8.

H9 *Calluna vulgaris – Deschampsia flexuosa* heath

Calluna vulgaris is almost always the most abundant plant in this community, often forming a fairly low and open canopy. Where burning is frequent, the individuals are immature and stands are uniform in age. No other sub-shrubs are consistently frequent throughout, although some can be quite common and locally abundant. *Vaccinium myrtillus* is the most important, particularly at higher altitudes. More locally *V. vitis-idaea* and *Empetrum nigrum* ssp. *nigrum* can be found. *Erica cinerea*, *E. tetralix* and *Ulex gallii* by contrast are very scarce.

The only other vascular constant is *Deschampsia flexuosa*, although even in open heather it often occurs only as sparse tufts, and under dense canopies it can almost disappear. *Molinia caerulea* can become frequent on moister ground, but *Agrostis capillaris*, *Holcus lanatus*, *H. mollis* and *Festuca rubra* only occur occasionally. Other herbs are also few and are of low cover. *Galium saxatile* and *Potentilla erecta* are frequent in grazed stands and *Juncus squarrosus* and *Pteridium aquilinum* occasionally occur. Seedlings of *Quercus* spp., *Betula* spp. and *Pinus sylvestris* may be seen but rarely survive to the sapling stage due to frequent burning and grazing.

The bryophyte and lichen flora is characteristic, although poor in species. *Hypnum cupressiforme s.l.* is restricted, but *Pohlia nutans* is constant and very common with occasional *Campylopus paradoxus* and *Dicranum scoparium*. *Ortho-dontium lineare* may be frequent. On exposed soil there can be locally abundant *Polytrichum juniperinum*, *P. piliferum* and *P. commune*. Among leafy hepatics *Gymnocolea inflata* is particularly characteristic. The commonest lichens are *Cladonia chlorophaea*, *C. floerkeana*, *C. squamosa*, *C. coniocraea* and *C. fimbriata*.

This heath is the characteristic sub-shrub vegetation of acid and impoverished soils at low to moderate altitudes through the Midlands and northern England. It is normally found on very base-poor soils with a surface acidity generally of pH 3-4, highly oligotrophic and at least moderately free-draining, often excessively so, which have been derived from a wide variety of parent materials. It is found mainly in the southern Pennines and North York Moors with more local occurrences scattered through the Midland plain.

The cool and wet climate has some influence on the floristics of this community, but much of its character derives from a combination of frequent burning and grazing. Also the heavy atmospheric pollution in the areas in which this heath occurs is thought to inhibit bryophyte and lichen diversity of the community. The community has been reduced considerably in extent. In the lowlands large tracts of heath have been reclaimed for agriculture whilst other areas have been lost to invasion by trees after the neglect of traditional treatments. Furthermore, both in the lowlands and around the upland fringes, the community has been replaced with coniferous plantations, or land use changes have led to the spread of U20 *Pteridium aquilinum – Galium saxatile* community.

H9

Vaccinium myrtillus and Campylopus paradoxus at most occasional, and other species listed opposite rare.

Vaccinium myrtillus or, more locally, *V. vitis-idaea. V. intermedium* or *Empetrum nigrum* common or locally abundant with occasional to frequent *Campylopus paradoxus, Gymnocolea inflata, Barbilophozia floerkii, Cladonia chlorophaea, C. floerkeana* and *C. squamosa.*

Molinia caerulea constant at low cover.

Molinia caerulea absent.

Deschampsia flexuosa often especially abundant with occasional *Holcus mollis* and *Festuca rubra; Galium saxatile* and *Potentilla erecta* frequent with occasional *Rumex acetosella.*

Deschampsia flexuosa may be abundant but not with associates listed opposite.

Hypnum cupressiforme and *Dicranum scoparium* common and sometimes abundant.

Calluna vulgaris and *Deschampsia flexuosa* often the only plants, with occasional *Pohlia nutans.*

H9b

Vaccinium myrtillus – *Cladonia* spp. sub-community

This is the richest sub-community characterised by younger canopies of heather, often recovering from burning. There are frequently one or more of the sub-shrubs listed above. Often *Deschampsia flexuosa* has a rather low cover. Among the sub-shrubs, bryophytes are more varied than in any other type of this heath. *Pohlia nutans, Campylopus paradoxus* and *Orthodontium lineare* all occur frequently and the leafy hepatics and lichens listed above are occasional to frequent.

This and the species-poor sub-community are the usual forms in the southern Pennines and the North York Moors and are widespread and sometimes extensive over heathlands that are still frequently burned.

H9e

Molinia caerulea sub-community

Calluna is generally very abundant, but *Deschampsia flexuosa* is frequently accompanied by small amounts of *Molinia caerulea.* The ground layer, however, is poorly developed with just very sparse *Pohlia nutans* and *Campylopus paradoxus.*

This sub-community is mainly found on wetter soils, and along with the *Galium* and *Hypnum* sub-communities, is primarily found on lowland sites where burning is no longer practised.

H9d

Galium saxatile sub-community

Calluna vulgaris remains constant but is often rivalled in cover by *Deschampsia flexuosa,* and stands are locally enriched by a little *Holcus mollis* or *Festuca rubra.* Commonly there are scattered plants or prominent patches of *Galium saxatile* and scattered *Potentilla erecta* with *Rumex acetosella* on bare areas. Lichens and hepatics are sparse and among the mosses only *Pohlia nutans* and *Hypnum cupressiforme s.l.* occur more than very occasionally.

This sub-community is mainly found on wetter soils, and along with the *Molinia* and *Hypnum* sub-communities, is primarily found on lowland sites where burning is no longer practised.

H9a

Hypnum cupressiforme sub-community

Calluna is typically strongly dominant but exceptionally the bushes tend to be large and mature or even degenerate. *Vaccinium myrtillus* and *Pteridium aquilinum* are both occasional and there are frequent, even dense tufts of *Deschampsia flexuosa.* The mosses are most distinctive; *Hypnum cupressiforme s.l.* is unusually common and abundant with *Dicranum scoparium* also preferential and frequently rivalling *Pohlia nutans* in its cover. Apart from occasional *Hypogymnia physodes* growing on older *Calluna vulgaris,* lichens are very few.

This sub-community is mainly found on wetter soils, and along with the *Galium* and *Molinia* sub-communities, is primarily found on lowland sites where burning is no longer practised.

H9c

Species-poor sub-community

In this, the most impoverished form, *Calluna vulgaris* and *Deschampsia flexuosa* are the only constants, and in frequently burned heather even the latter can almost disappear. *Vaccinium myrtillus* is occasional. *Pohlia nutans, Campylopus paradoxus* and *Orthodontium lineare* all show reduced frequencies compared with H9b.

This and the *Vaccinium* sub-community are the usual forms in the southern Pennines and the North York Moors, and are widespread and sometimes extensive over heathlands that are still frequently burned.

H10 *Calluna vulgaris – Erica cinerea* heath

This community is typically dominated by *Calluna vulgaris*, but the cover, height and structure of the sub-shrub canopy vary markedly depending on the intensity and timing of burning and grazing. *Erica cinerea*, a constant, is frequent but generally subordinate to heather and persists below taller *Calluna* canopies. *Vaccinium myrtillus*, by contrast, is at most occasional and *V. vitis-idaea* is scarce. *Empetrum nigrum* ssp. *nigrum* can occur, but mainly in sub-community H10b. The restricted occurrence of these sub-shrubs is a contrast with *Calluna vulgaris – Vaccinium myrtillus* heath (H12).

Apart from the abundance of the two constant sub-shrubs there are two other distinctive floristic features of this type of heath. These are firstly the high frequency of grasses and to a lesser extent sedges and dicotyledons, and secondly the striking contribution that the ground layer makes to this community. *Deschampsia flexuosa* is the most consistent grass throughout, with *Agrostis canina* and *Nardus stricta* occasional to frequent. In certain sub-communities *Festuca ovina*, *Anthoxanthum odoratum*, *Agrostis capillaris* and *Molinia caerulea* become very common. *Carex binervis* and *C. pilulifera* are very characteristic of this community. After burning, mixtures of these plants can become patchily abundant and *Deschampsia flexuosa* and *C. pilulifera* temporarily dominant. There are typically only a few dicotyledons, but *Potentilla erecta* is a constant and *Galium saxatile* is fairly common.

After burning, a local abundance of *Polytrichum piliferum*, *P. juniperinum* and encrusting *Cladonia* species can develop. In exposed stands there is often a patchy carpet of *Racomitrium lanuginosum* and fruticose lichens. However, more important than these species in the community as a whole are bulky pleurocarpous mosses such as *Hypnum cupressiforme s.l.*, *Pleurozium schreberi* and *Hylocomium splendens*, with *Rhytidiadelphus triquetrus* and *R. loreus* also occurring occasionally. These species, with *Dicranum scoparium*, become abundant with the maturing and opening up of the *Calluna* bushes.

This heath is characteristic of acid to circum-neutral and generally free-draining soils in the cool oceanic lowlands and upland fringes of northern and western Britain. The soils on which this community is found can be quite moist as a result of the climate and the superficial pH beneath the community can be anywhere between 3.5 and 6. It occurs widely through the more oceanic parts of Scotland, with outlying stands in Wales, western England and around the east-central Highlands.

In more exposed situations it may be considered as an edaphic or climatic climax, but often burning and grazing are important in controlling its composition and structure. Steady grazing pressure pushes the vegetation towards the *Festuca ovina – Agrostis capillaris – Galium saxatile* grassland (U4) or, over more base-rich soils, the *Festuca ovina – Agrostis capillaris – Thymus praecox* grassland (CG10). After fire, heavy grazing can precipitate a run-down of the heath to swards in which *Nardus stricta* or *Juncus squarrosus* play an important part or permit the spread of *Pteridium aquilinum*. Release from grazing and burning, in all but the most exposed sites, would theoretically permit progression to scrub and woodland, although in many areas natural seed parents are now scarce.

H10

Sub-shrub canopy often short in a grassy heath with frequent *Festuca ovina, F. rubra, Agrostis capillaris, Anthoxanthum odoratum* and *Galium saxatile* and occasional *Campanula rotundifolia, Succisa pratensis* and *Hypericum pulchrum. Dicranum scoparium, Pleurozium schreberi* and *Hylocomium splendens* patchy.

Empetrum nigrum ssp. *nigrum* and *Scirpus cespitosus* occasional at most and *Molinia caerulea* very common at low covers with frequent *Carex binervis* and occasional *Juncus squarrosus. Campylopus paradoxus, Sphagnum capillifolium* and *Diplophyllum albicans* occasional to frequent.

Empetrum nigrum ssp. *nigrum* quite common and *Scirpus cespitosus* patchily prominent with frequent *Carex panicea* and *C. pilulifera* and occasional *Huperzia selago. Racomitrium lanuginosum* common and often abundant among degenerating bushes with patches of *Cladonia uncialis* and *C. impexa.*

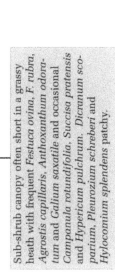

Danthonia decumbens very common with occasional to frequent *Carex pulicaris, Viola riviniana, Linum catharticum, Prunella vulgaris* and *Primula vulgaris.*

Combinations of such species rare.

H10a

Typical sub-community

In this, the most species-poor sub-community, *Calluna* is typically dominant and abundant in pioneer and building regrowth after burning. *Erica cinerea* is very frequent and can be prominent. *Vaccinium myrtillus* is occasional and *Empetrum nigrum nigrum* and *Erica tetralix* scarce. Monocotyledons are few with *Deschampsia flexuosa* very frequent and sometimes prominent. *Molinia caerulea* is preferential and patchily abundant with occasional *Scirpus cespitosus* and *Juncus squarrosus. Carex binervis* is well represented. Apart from *Potentilla erecta* and *Galium saxatile*, dicotyledons are very sparse. The ground layer is also poor in species and of low cover.

This sub-community is found throughout the range of H10.

H10b

Racomitrium lanuginosum sub-community

This is found on exposed sites where the sub-shrub canopy is more open and *Calluna vulgaris* is the usual dominant. *Erica cinerea* is frequent, *Vaccinium myrtillus* occasional and *Empetrum nigrum* spp. *nigrum* is preferential and quite common. Grasses are sparse with scattered tufts of *Deschampsia flexuosa* other grasses very occasional. *Carex binervis* is rare, its place being taken by *Carex pilulifera* and *C. panicea. Scirpus cespitosus* is also frequent. *Potentilla erecta* is the only frequent dicotyledon *Huperzia selago* is preferentially common. There are substantial areas of the ground layer with *Racomitrium lanuginosum* the most abundant moss, and frequent *Hypnum cupressiforme* s.l. Lichens are well represented.

This sub-community is typical of the Western Isles and Shetland.

H10c

Festuca ovina – Anthoxanthum odoratum sub-community

Calluna vulgaris is still abundant but *Erica cinerea* may often be co-dominant. The sub-shrubs are usually short, commonly forming a mosaic with a grassy turf. Most frequent here are the grasses and other species listed above. *Carex binervis* and *C. pilulifera* are also common. Dicotyledonous herbs are more numerous than in 10a and 10b. *Potentilla erecta* and *Galium saxatile* are both very common with occasional records for several species . Bulky pleurocarpous mosses are consistent and distinctive here with frequent *Hypnum cupressiforme s.l., Pleurozium schreberi, Hylocomium splendens* and also *Dicranum scoparium.*

This sub-community is common in south-west Scotland.

H10d

Thymus praecox – Carex pulicaris sub-community

This heath is found on relatively base-rich brown earth soils and is very similar to H10c with *Calluna vulgaris* and *Erica cinerea* both able to show prominence and with herbs and bryophytes both being of structural importance. Here there are additional preferentials, making this the most species-rich sub-community. The species listed above are most frequent together with *Carex panicea* and *Thymus praecox.* Among the bryophytes *Dicranum scoparium* and the pleurocarps remain very common; additionally *Rhytidiadelphus triquetrus* and *Breutelia chrysocoma* are frequent.

This sub-community is local in occurrence but can be found on Skye, Rum and Uist and scattered localities through the Highlands and Southern Uplands.

H11 *Calluna vulgaris – Carex arenaria* heath

Calluna vulgaris is the only constant sub-shrub found in this community and is often abundant, although cover may be discontinuous and patchy in younger or grazed stands. Other frequent sub-shrubs are *Erica cinerea* and *Empetrum nigrum* ssp. *nigrum*, and each can be locally abundant, to the exclusion of *Calluna* itself. Sometimes *Rosa pimpinellifolia* is plentiful, and together with *Erica tetralix* and *Salix repens* is found in transitions to wetter heath.

Carex arenaria is constant, but no more than moderately abundant and often senile, except where the sand is locally mobile. *Ammophila arenaria* is also frequent throughout, though usually sparse. In more species-poor stands these may be the only species, but often there is some *Festuca rubra* (or *F. ovina*) with *Agrostis capillaris* and *Anthoxanthum odoratum* and less commonly *Poa pratensis*. Variation among dicotyledons is modest, but *Galium verum*, *Lotus corniculatus*, *Viola riviniana* and *Thymus praecox* all occur quite frequently with several other herbs.

There may be hypnoid mosses such as *Hypnum cupressiforme s.l.*, *Pleurozium schreberi*, *Hylocomium splendens* and *Rhytidiadelphus triquetrus* in the turf. On areas of bare ground acrocarps such as *Polytrichum juniperinum*, *P. piliferum* and *Ceratodon purpureus* may be patchily abundant.

This is the characteristic sub-shrub vegetation of stabilised, base-poor sands on dunes and plains around the coasts of Britain. The heath is largely confined to sands with a pH of less than 5 and can only establish on sediments with surface stability such as found on older dunes and on consolidated sand plains. It is very local along the coasts of western England and Wales, becoming commoner in Scotland.

The community develops in primary succession by colonising fixed dune grasslands on acid sands or where more lime-rich sands have become leached. Relief from grazing is probably important for its establishment but once established predation by herbivores, along with variation in regional climate and substrate, influences its composition and structure, and ultimately, grazing maintains the community against reversion to grassland or progression to scrub and woodland.

H11

Erica cinerea constant and *Aira praecox* frequent but *Empetrum nigrum nigrum*, *Agrostis capillaris* and *Galium saxatile* are rare. *Dicranum scoparium* common and some of *Cladonia furcata*, *C. floerkiana*, *C. pyxidata*, *C. gracilis* and *C. foliacea* in an often extensive carpet of lichens.

Erica cinerea and *Aira praecox* only occasional at most, but *Empetrum nigrum nigrum*, *Agrostis capillaris* and *Galium saxatile* very frequent. *Cladonia arbuscula* and *C. impexa* can be patchily prominent but stretches of lichen-rich turf are not characteristic.

The associates listed opposite are all scarce in impoverished mixtures of *Calluna vulgaris* and *Carex arenaria*.

H11a

Erica cinerea sub-community

This sub-community is prominent in southern areas where rainfall is relatively low. *Calluna vulgaris* and *Erica cinerea* are generally co-dominant with *Erica cinerea* sometimes colonising first. *Rosa pimpinellifolia* is a distinctive invader. *Festuca rubra/ovina* is very common with scattered shoots of *Carex arenaria* and *Luzula campestris*. Other herbs are occasional. The distinctive element is the cryptogams, often occupying the bulk of the ground. Hypnoid mosses are scarce and most obvious are the lichens, particularly *Cladonia* spp., including those listed above. *Cornicularia aculeata* and *Hypogymnia physodes* are also common, sometimes with *Peltigera canina* and low-growing *Usnea* species.

This is the most widespread sub-community but it is replaced locally by the *Empetrum* sub-community in north and east Scotland.

H11b

Empetrum nigrum ssp. nigrum sub-community

Calluna vulgaris and *Empetrum nigrum*, which forms large patches, are the co-dominants here, especially in the cooler wetter north of Britain. *Festuca rubra/ovina* remains common among frequent *Carex arenaria* and *Luzula campestris* with other species as listed above. Hypnoid mosses are more prominent with *Hypnum cupressiforme*, *Pleurozium schreberi*, *Rhytidiadelphus triquetrus* and *Hylocomium splendens* and less commonly *Ptilidium ciliare*. Lichens are more patchy.

This sub-community is found locally in north and east Scotland.

H11c

Species-poor sub-community

Calluna vulgaris is overwhelmingly dominant with only occasional or even no bushes of other species. *Carex arenaria* remains constant but grasses may only be represented by a few tufts of *Anthoxanthum odoratum* or *Deschampsia flexuosa*. Other herbs are also sparse. Bryophytes may include hypnoid species and *Dicranum scoparium*.

This sub-community can be found throughout the range of H11.

H12 *Calluna vulgaris –*
Vaccinium myrtillus heath

This heath is generally dominated by *Calluna vulgaris* and includes most of the Calluneta from less oceanic sub-montane areas where burning is commonly practised, including many grouse moors. Here a predominance of building-phase *Calluna* is found, but a more open cover of degenerate *Calluna* can often also be present. *Vaccinium myrtillus* is constant though it is usually subordinate to *Calluna* and is most vigorous out of reach of grazing animals. *Vaccinium vitis-idaea* is also found, sometimes with local prominence, and *Erica cinerea* may also be present on drier slopes. *Empetrum nigrum* ssp. *nigrum* is frequent, forming mats after burning, but then becomes reduced after the *Calluna* has regrown.

In many stands herbs are rare. Only *Deschampsia flexuosa* is frequent throughout. When grazing is regular there may be additional herbs including *Festuca ovina*, *Agrostis capillaris*, *A. canina*, *Nardus stricta*, *Potentilla erecta* and *Galium saxatile*.

The ground-layer is often prominent with bulky mosses characteristic, such as *Dicranum scoparium*, *Pleurozium schreberi*, *Hypnum cupressiforme s.l.* and *Hylocomium splendens*, together with larger *Cladonia* species. Encrusting lichens and *Polytrichum* species can be abundant in the years following burning.

This community is the typical sub-shrub community of acidic to circumneutral, free-draining mineral soils throughout the cold and wet sub-montane zone generally between 200 m and 600 m. The soils on which it occurs are widespread throughout this zone, developing from a variety of siliceous parent materials, intrusive igneous rock or coarse glacio-fluvial gravels. Despite being free-draining the soils are normally moist for the majority of the year because of the climate and the superficial pH is usually between 3.5 and 4.5. It is extensive in the east-central Highlands but also important in south-east Scotland, the Lake District, parts of Wales and the South-West Peninsula and the North York Moors. In places like the southern Pennines, where air pollution is severe, it is largely replaced by *Calluna vulgaris – Deschampsia flexuosa* heath (H9).

Burning and grazing are the major influences on floristics and structure, although climatic and edaphic difference play some part in determining variation within the community. Successional developments are usually held in check by burning and grazing and without these most stands would probably progress to scrub and woodland. Continuous heavy grazing favours the loss of sub-shrub vegetation to grassland and in some instances, particularly after burning, may result in the spread of *Pteridium aquilinum*.

H12

Vaccinium vitis-idaea and *Empetrum nigrum* both scarce in rather species-poor heath, usually overwhelmingly dominated by *Calluna vulgaris*.

H12a

Calluna vulgaris sub-community

Vegetation is typically species-poor with *Calluna vulgaris* overwhelmingly dominant and other sub-shrubs of low cover. *Vaccinium myrtillus* is very frequent and *Erica cinerea* common, but both only as scattered shoots. Other vascular associates are few. *Deschampsia flexuosa* is frequent as scattered shoots and sparse plants of *Potentilla erecta* and *Pteridium aquilinum* are quite common. The ground cover is not extensive and only *Dicranum scoparium*, *Hypnum jutlandicum* and *Pleurozium schreberi* occur frequently as scattered shoots.

Vaccinium vitis-idaea and *Empetrum nigrum* frequent, with occasional *Juncus squarrosus* and *Blechnum spicant*.

Sub-shrub cover extensive and varied with *Vaccinium vitis-idaea* especially frequent and species listed opposite typically scarce. Cryptogam flora varied and often abundant, with bulky pleurocarps prominent, *Hylocomium splendens* often joining *Pleurozium schreberi* and *Hypnum jutlandicum*. *Cladonia impexa*, *C. uncialis* and *C. pyxidata* common.

H12b

Vaccinium vitis-idaea – Cladonia impexa sub-community

This includes most of the richer stands of this heath, which develop a number of years after burning. Although *Calluna vulgaris* is still the general dominant it is frequently accompanied by *Vaccinium myrtillus*, *V. vitis-idaea* and *Empetrum nigrum* and occasionally with *Erica cinerea*. Herbs are generally sparse with only scattered plants of *Deschampsia flexuosa*, and occasional *Potentilla erecta*, *Juncus squarrosus* and *Blechnum spicant*. Bryophytes and lichens are more numerous including the species listed above.

Calluna vulgaris not so abundant as usual, and other sub-shrubs of generally moderate cover in a grassy heath, with frequent *Festuca ovina*, *Agrostis capillaris*, *Nardus stricta*, *Galium saxatile* and *Potentilla erecta* and occasional *Carex pilulifera*, *Campanula rotundifolia* and *Polygala serpyllifolia*. Lichens not extensive, often patchy.

H12c

Galium saxatile – Festuca ovina sub-community

This sub-community is found on better soils and after burning, often followed by grazing. *Calluna vulgaris* is less dominant and with other sub-shrubs forms an open growth within a grassy sward. *Deschampsia flexuosa* is joined by a variety of herbs including those listed above. Where the soils are less base-poor, species such as *Lotus corniculatus*, *Lathyrus montanus*, *Succisa pratensis*, *Viola riviniana* and *Anemone nemorosa* can be locally abundant. Bryophytes remain quite varied, but lichens are few and of low cover.

H13 *Calluna vulgaris – Cladonia arbuscula* heath

This heath has a dwarfed mat of sub-shrubs with few vascular associates, but with a prominent lichen flora. *Calluna vulgaris* is the most frequent species, generally prostrate and forming a carpet or in wave-like bands or on solifluction terraces. Among other sub-shrubs *Empetrum nigrum* is most important, usually as ssp. *hermaphroditum*, but with ssp. *nigrum* at lower altitudes. It may be intermixed in the mat or forming clumps. *Loiseleuria procumbens* is quite frequent and abundant, but *Arctostaphylos uva-ursi* is at most occasional. Both *Vaccinium myrtillus* and *V. vitis-idaea* are common, but always subordinate in cover.

The other vascular associates are few and sparse. *Deschampsia flexuosa* and *Carex bigelowii* are most frequent with species such as *Scirpus cespitosus*, *Agrostis canina* and *Molinia caerulea* at lower altitudes and *Juncus trifidus* becoming occasional at higher levels. *Huperzia selago* is also frequent in higher altitude stands.

Lichens are important structurally. *Cladonia arbuscula* is especially common and, where there is some shelter, may be abundant. It is usually mixed with *C. rangiferina* which locally may be co-dominant. Also constant are *C. uncialis*, *Cetraria islandica*, *Alectoria nigricans* and *Cornicularia aculeata*. Among these, bryophytes are generally few and rarely of any abundance. *Racomitrium lanuginosum* is constant and can form locally conspicuous patches.

This heath is the characteristic sub-shrub vegetation of base-poor soils, over exposed ridges and summits of mountains, in parts of Britain with a cold continental climate. It is found on soils with a superficial pH of between 4 and 5, and frequently a humic surface above pervious acidic bedrocks and superficials. It is most widespread through the east-central Highlands of Scotland, thinning out westwards into the central Grampians and north-west Highlands where it is progressively replaced by its oceanic counterpart *Calluna vulgaris – Racomitrium lanuginosum* heath (H14). There are a few fragmentary localities in northern England and Wales.

It is a vegetation type of unsheltered slopes generally between 600 m and 900 m where there are almost constant strong winds which frequently clear the ground of snow and subject the vegetation not only to reduced precipitation but also to the effects of frequent and severe frosts and subsequent thaws. Burning and grazing may have curtailed its range in suitable localities in the more southerly uplands, but in the eastern Highlands the vegetation seems to be largely unaffected by treatments and the community can be considered a climax.

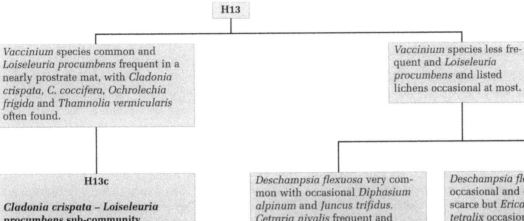

H13

Vaccinium species common and *Loiseleuria procumbens* frequent in a nearly prostrate mat, with *Cladonia crispata, C. coccifera, Ochrolechia frigida* and *Thamnolia vermicularis* often found.

Vaccinium species less frequent and *Loiseleuria procumbens* and listed lichens occasional at most.

H13c

***Cladonia crispata – Loiseleuria procumbens* sub-community**

Calluna vulgaris is again the usual dominant, but one or both subspecies of *Empetrum nigrum* are common and can be abundant. In the flattened sub-shrub mat there is a mixed carpet of lichens including the species listed above. The herbs comprise scattered shoots of *Carex bigelowii, Huperzia selago* and *Deschampsia flexuosa*.

Deschampsia flexuosa very common with occasional *Diphasium alpinum* and *Juncus trifidus*. *Cetraria nivalis* frequent and *Cladonia gracilis* and *C. bellidiflora* occasional.

Deschampsia flexuosa only occasional and *Carex bigelowii* scarce but *Erica cinerea* and *E. tetralix* occasional among the sub-shrubs, sometimes with *Scirpus cespitosus* and *Molinia caerulea*. *Cetraria nivalis* and *Cladonia gracilis* uncommon, but *C. arbuscula* and *C. rangiferina* especially abundant.

H13b

***Empetrum nigrum* ssp. *hermaphroditum – Cetraria nivalis* sub-community**

On bleak exposed sites at higher altitudes lichens remain abundant, but *Calluna vulgaris* or occasionally *Empetrum nigrum hermaphroditum* is dominant. Both *Vaccinium myrtillus* and *V. vitisidaea* are more frequent than in H13a but usually have low cover. *Loiseleuria procumbens* is occasional. *Cladonia arbuscula* is still the most frequent lichen but is commonly joined by the species listed above. Mosses only make a minor contribution and the few vascular associates are present as scattered individuals.

H13a

***Cladonia arbuscula – Cladonia rangiferina* sub-community**

This sub-community is found at lower altitudes or more sheltered sites where larger *Cladonia* species, as above, are especially abundant, often exceeding the sub-shrubs in cover. *Cladonia impexa* can also be found, but *C. gracilis* and *C. crispata* are scarce. The most common sub-shrub is *Calluna vulgaris* but *Empetrum nigrum* is very common, often as ssp. *nigrum. Vaccinium* species are only occasional.

101

H14 *Calluna vulgaris – Racomitrium lanuginosum* heath

This heath consists essentially of a dwarfed sub-shrub mat with *Calluna vulgaris* usually predominant, together with *Racomitrium lanuginosum.* Other sub-shrubs play a subordinate role, but may be common. Most frequent is *Empetrum nigrum*, with the two subspecies characterising opposite ends of the altitudinal range (ssp. *nigrum* preferentially common towards lower levels and ssp. *hermaphroditum* largely confined to higher altitudes). *Erica cinerea* is also frequent.

Other vascular associates are few and usually scattered. *Deschampsia flexuosa*, *Huperzia selago*, *Carex pilulifera*, *Potentilla erecta* and *Scirpus cespitosus* are all frequent, and *Carex bigelowii* becomes common at higher altitudes.

The extensive woolly carpet of *Racomitrium lanuginosum* which can be up to 5-10 cm thick is the most noticeable feature of this community. *Hypnum cupressiforme s.l.* is also very frequent in some stands, often with several other mosses and occasional hepatics. Lichens are common and varied but not abundant, and species like *Cetraria nivalis* and *Alectoria ochroleuca* are absent. *Cladonia arbuscula* and *C. uncialis* are the most frequent, and *Sphaerophorus globosus* and *Cornicularia aculeata* are also common throughout. *Cladonia impexa* is frequent at lower altitudes and *Cladonia gracilis*, *C. bellidiflora*, *Cetraria islandica* and *Ochrolechia frigida* occur often at higher altitudes.

This community is the typical sub-shrub community of base-poor soils on windswept plateaux and ridges at moderate to fairly high altitudes in the cool oceanic climate of the mountains of north-west Scotland. It can be found up to 750 m, although this can extend up to 1000 m in the east; to the west and north, on islands like Skye, Orkney and Shetland, it can extend down to below 250 m. The community is found on the base-poor rankers and podzolic soils which are widespread in this region, with a superficial pH between 4 and 5 and a humic surface. It is very much a community of the north-west Highlands with scattered occurrences in the central Grampians.

Like its eastern counterpart *Calluna vulgaris – Cladonia arbuscula* heath (H13) it is found over gentle to moderately steep slopes which are exposed to fairly constant strong winds that clear the snow which might otherwise provide shelter in the coldest months. Although it is sometimes grazed by sheep and deer, it is unlikely that this factor is important in maintaining the characteristic composition and physiognomy, and this vegetation can be regarded as the natural climax in such exposed situations in its range.

H14

Arctostaphylos uva-ursi constant with frequent *A. alpinus*, *Empetrum nigrum nigrum* and *Erica cinerea*. *Scirpus cespitosus* common with *Molinia caerulea* and *Carex binervis* occasional. *Dicranum scoparium*, *Diplophyllum albicans* and *Pleurozium schreberi* frequent.

Empetrum nigrum nigrum and *Erica cinerea* at most occasional. *Arctostaphylos* species and the herbs and bryophytes listed opposite very scarce.

Empetrum nigrum hermaphroditum and herbs and lichens listed opposite occasional at most. *Festuca ovina* and *Agrostis canina* very common, sometimes with *Antennaria dioica*, *Carex panicea*, *Thymus praecox* and *Euphrasia micrantha*.

Empetrum nigrum hermaphroditum constant with frequent *Nardus stricta* and *Diphasium alpinum* and sometimes with an extensive lichen cover, including *Cetraria islandica*, *Cladonia gracilis* and *Ochrolechia frigida*.

H14a

Festuca ovina sub-community

Calluna vulgaris or *Racomitrium lanuginosum* or mixtures of the two dominate the vegetation mat with other species playing only a minor role. *Empetrum nigrum* ssp. *nigrum* and *hermaphroditum* and *Erica cinerea* are only occasional and in some stands *Loiseleuria procumbens* can be prominent. More striking is the variety of herbaceous associates. *Deschampsia flexuosa* is less common than usual, but *Carex bigelowii* is frequent with *C. pilulifera*, *Huperzia selago* and *Potentilla erecta*. More preferential are the species listed above. Lichen cover is comparatively low but *Cladonia uncialis* is very common.

This and the *Arctostaphylos* sub-community are found at the lowest altitudes and most sheltered sites occupied by the community.

H14b

Empetrum nigrum ssp. hermaphroditum sub-community

Calluna vulgaris generally abundant with *Racomitrium lanuginosum* sometimes co-dominant but often subordinate. The variety of herbs characteristic of H14a is not found although *Deschampsia flexuosa* and *Carex bigelowii* are very frequent and *Potentilla erecta*, *Huperzia selago* and *Carex pilulifera* remain common. Most distinctive are the cryptogams with several pleurocarpous mosses and the typical lichen flora of the community with additional species, including those listed above.

This is the typical form of this sub-community and is found at the highest altitudes of the range of H14.

H14c

Arctostaphylos uva-ursi sub-community

Calluna vulgaris and *Racomitrium lanuginosum* retain the representation of H14b but the sub-shrubs are more varied with *Erica cinerea* and *Empetrum nigrum* ssp. *nigrum* at their peak of frequency, and locally abundant *Arctostaphylos uva-ursi* and *A. alpinus*. Among the vascular associates *Carex bigelowii* is very scarce. In the bryophyte mat *Dicranum scoparium* and larger pleurocarps such as *Hypnum cupressiforme* s.l. and *Pleurozium schreberi* make their biggest contribution. In the lichen flora *Cladonia impexa* is common together with the community species.

This and the *Festuca* sub-community are found at the lowest altitudes and most sheltered sites occupied by the community.

H15 *Calluna vulgaris – Juniperus communis* ssp. *nana* heath

Prostrate juniper, referable to *Juniperus communis* ssp. *nana*, is occasional in a wide variety of sub-shrub heaths. Here, however, it is consistently dominant in the sub-shrub mat, accompanied by a small but distinctive group of oceanic hepatics. The mat is generally less than 10 cm high, fairly continuous in the best stands, but it may form a mosaic with islands of vegetation on tracts of bare rock and debris. Several other sub-shrubs are well represented: *Calluna vulgaris* and *Erica cinerea* are especially frequent and the former often fairly abundant. *Arctostaphylos uva-ursi* and *A. alpinus* are less common, and *Empetrum nigrum* ssp. *hermaphroditum* is occasional.

Vascular associates are typically few and are usually scattered in the mat. *Deschampsia flexuosa*, *Scirpus cespitosus* and *Potentilla erecta* are constant, with *Huperzia selago*, *Solidago virgaurea*, *Dactylorhiza maculata*, *Polygala serpyllifolia*, *Succisa pratensis*, and *Antennaria dioica* more occasional.

In some stands the cryptogam flora is similar to other kinds of dwarfed sub-shrub heath. In typical examples of this community, however, the species *Racomitrium lanuginosum*, *Cladonia uncialis*, *C. impexa*, *Sphaerophorus globosus* and *Cornicularia aculeata*, which are common in all these other kinds of heath, are joined by *Pleurozia purpurea*, *Frullania tamarisci* and *Diplophyllum albicans* which are not. Where the sub-shrub canopy is well-developed the total cover of the cryptogams is much less than in the typical moss-heaths of the region.

This heath is confined to humic rankers at moderate altitudes in the cool oceanic climate along the western seaboard of the north-west Highlands and some of the Western Isles. Soil development under this community is typically rudimentary with just shallow accumulations of decaying juniper and bryophyte litter on Cambrian quartzite screes. Although perhaps once more widespread throughout the north-west Highlands, the community is now of rather patchy occurrence along the western side of the more northerly mountains with especially good stands on Beinn Eighe and Foinaven. The community is replaced in the continental climate of the east-central Highlands by the *Juniperus communis* ssp. *communis – Oxalis acetosella* woodland (W19).

It is confined to the lower portion of the altitudinal ranges of the other dwarf sub-shrub heaths and, although the vegetation mat is typically blown clear of snow, is not usually found in the kind of severely exposed situations of which the other communities are so characteristic. This community is given some protection against the effects of grazing by the rocky ground on which it is typically found, but it is readily damaged by burning.

No sub-communities.

H16 *Calluna vulgaris – Arctostaphylos uva-ursi* heath

Although *Arctostaphylos uva-ursi* is found as an occasional in a variety of heath types, it is most often found in this community, which has a distinct boreal character. *Calluna vulgaris* is always present and is the most usual dominant, forming a canopy 20-40 cm high and having a substantial total cover. *Arctostaphylos uva-ursi* is constant and can become modestly abundant in gaps within the heather cover. *Erica cinerea* is also very common but of low cover. In many stands there is some *Vaccinium myrtillus* and *V. vitis-idaea*.

Quite commonly there are small amounts of *Genista anglica*, but herbaceous associates are few except in the *Pyrola media – Lathyrus montanus* sub-community. The only constant grass is *Deschampsia flexuosa* and this can be joined by *Luzula multiflora* and *L. pilosa*.

Bryophytes are variable, with the bulkier mosses often strongly associated with particular stages in the heather regeneration cycle. *Hypnum jutlandicum*, *Pleurozium schreberi* and *Dicranum scoparium*, however, are very common overall and *Hylocomium splendens* is also a constant through much of the community.

Lichens also differ in their representation, with only *Cladonia impexa* constant and, in many stands, of low cover. Fruticose species such as *C. arbuscula* and *C. rangiferina* tend to follow the larger pleurocarps in developing among the more shady and humid conditions of older heather canopies. *Hypogymnia physodes* and, less commonly, *Cetraria glauca* can be seen on decaying woody stems.

This heath is characteristic of base-poor to circumneutral soils at moderate altitudes, generally between 250 m and 600 m altitude, in the cold continental climate of the east-central Highlands of Scotland. It is found on a variety of acid soils developed from lime-poor parent material. It occurs widely but fairly locally through the east-central Highlands with especially good representation in Speyside.

The community forms an important part of grouse-moor in the central Highlands and although edaphic differences play some part in determining floristic variation in the community, their effects are often overlain and modified by the influence of burning which ultimately maintains this vegetation as a plagioclimax. Stretches of moorland including stands of the community are often open to livestock but there is little information on the impact of grazing on this vegetation.

H16

Arctostaphylos uva-ursi often extensive with Calluna vulgaris, Erica cinerea and Vaccinium species. Festuca ovina common with occasional Agrostis capillaris and Anthoxanthum odoratum. The rich herb flora includes frequent Potentilla erecta, Pyrola media, Viola riviniana, Lathyrus montanus, Hypericum pulchrum, Anemone nemorosa, Trientalis europaea, Galium saxatile and Lotus corniculatus.

H16a

Pyrola media – Lathyrus montanus sub-community

Arctostaphylos uva-ursi is most prominent in this sub-community, on less acidic brown earth soils, forming a patchwork with Calluna vulgaris. Erica cinerea can have a high cover and the two Vaccinium species are common. Genista anglica is fairly frequent. The most striking feature is the associated herb flora developed in more mesotrophic conditions. Grasses are more common with Festuca ovina frequently joining Deschampsia flexuosa together with the herbs listed above. The mosses characteristic of H16 are common, but not abundant, including the preferential Rhytidiadelphus triquetrus.

Calluna vulgaris usually a strong dominant with Arctostaphylos uva-ursi subordinate and associates listed opposite occasional at most.

Vaccinium myrtillus commonly accompanies V. vitis-idaea in a sparse understorey with Empetrum nigrum intermingled. Hypnoid mosses often extensive with Cladonia arbuscula and C. rangiferina also frequent.

H16b

Vaccinium myrtillus – Vaccinium vitis-idaea sub-community

Arctostaphylos uva-ursi remains constant here but Calluna vulgaris is usually dominant and with the sub-shrubs mentioned above. In contrast with H16a, herbs are scarce though Festuca ovina, Carex pilulifera, Potentilla erecta, Luzula multiflora, L. pilosa and Listera cordata are occasional. Cryptogams, however are more diverse and extensive with the bulky pleurocarps Hypnum jutlandicum, Pleurozium schreberi and Hylocomium splendens often abundant. Larger lichens are also more apparent including several Cladonia species.

Vaccinium myrtillus absent, V. vitis-idaea and Empetrum nigrum nigrum scarce. and hypnoid mosses patchy. Scirpus cespitosus and Carex pilulifera common and lichens extensive on areas of bare ground with frequent Cladonia uncialis, C. impexa, C. floerkeana, C. coccifera and C. squamosa.

H16c

Cladonia spp. sub-community

Although Arctostaphylos uva-ursi is sometimes quite abundant, Calluna vulgaris is more often overwhelmingly dominant. Both Erica cinerea and Genista anglica occur frequently. The herbs of sub-community H16a are hardly ever found. Among the bryophytes the characteristic mosses Hylocomium splendens and Pleurozium schreberi are very patchy, but Hypnum jutlandicum and Dicranum scoparium remain frequent often with a little Pohlia nutans. Peatencrusting lichens are most noticeable, particularly the Cladonia species listed above.

H17 *Calluna vulgaris – Arctostaphylos alpinus* heath

Arctostaphylos alpinus occurs with some frequency in various kinds of dwarfed sub-shrub heath, but is most typical of this community where it is a constant, although usually a subordinate one, in the woody mat which is normally less than 10 cm tall. It is usually dominated by stunted bushes of *Calluna vulgaris* with stretches of bare stones between. *Empetrum nigrum* ssp. *hermaphroditum* is strongly preferential to higher altitudes and ssp. *nigrum* is largely confined to lower situations. *Loiseleuria procumbens* is characteristically found with *E. nigrum* ssp. *hermaphroditum*, and *Erica cinerea* with *E. nigrum* ssp. *nigrum*. *Vaccinium myrtillus* is common throughout but other *Vaccinium* species are scarce.

There are few herbs. *Huperzia selago* is the commonest and a constant, and is often accompanied, at higher altitudes, by *Diphasium alpinum*, *Carex bigelowii* and *Antennaria dioica*. *Deschampsia flexuosa* is also frequent throughout, though more so at lower altitudes where *Potentilla erecta*, *Scirpus cespitosus* and *Carex pilulifera* occur most commonly.

More conspicuous are the lichens which form a patchy mosaic. *Cladonia arbuscula* and *C. uncialis* are constant. Preferential to higher altitudes are *Cetraria glauca*, *C. islandica*, *Cornicularia aculeata*, *Alectoria nigricans* and *Sphaerophorus globosus*. Mosses are not abundant. *Racomitrium lanuginosum* is constant, though in small amounts, and *Hypnum jutlandicum* becomes frequent at lower altitudes.

This heath is the typical climax sub-shrub vegetation of rather base-poor moder soils over very exposed ridges and crests at moderate to fairly high altitudes in the cold and humid climate of the mountains of north-west Scotland. It is found at higher altitudes than the *Calluna vulgaris – Racomitrium lanuginosum* heath (H14) which has a similar distribution, and its normal range is between 500 m and 750 m, although it can exceptionally be found up to 900 m, and down to 250 m along the north Scottish coast and on Orkney. It is typically found on humic poor rankers and more occasionally mature podzolised soil that have been derived from a variety of lime-poor parent materials. This community is confined to the north-west Highlands, the north Scottish coast and Orkney.

The community may be lightly grazed by sheep and deer but this probably has little effect on its floristics or physiognomy. The inhospitable environment and the harsh conditions maintain the vegetation as a climax. Burning is very deleterious and may cause damage from which recovery is extremely slow if not impossible. Burning may have eliminated this community from many sites throughout its range.

H17

Empetrum nigrum ssp. *hermaphroditum* and *Loiseleuria procumbens* constant with frequent *Carex bigelowii*, *Diphasium alpinum*, *Antennaria dioica*, *Cladonia uncialis*, *C. arbuscula*, *C. gracilis*, *C. pyxidata*, *C. bellidiflora*, *Cetraria glauca*, *C. islandica*, *Alectoria nigricans* and *Sphaerophorus globosus*.

Loiseleuria procumbens occasional, but *Empetrum nigrum* ssp. *hermaphroditum* replaced by ssp. *nigrum* and with *Erica cinerea* becoming common. *Potentilla erecta* very frequent with *Scirpus cespitosus* and *stricta* often present. *Cladonia uncialis* and *C. arbuscula* remain common but lichen flora not so varied or abundant.

H17a

***Loiseleuria procumbens – Cetraria glauca* sub-community**

In this distinctive sub-community mixtures of *Calluna vulgaris* with subordinate *Arctostaphylos alpinus*, *Loiseleuria procumbens* and *Empetrum nigrum* ssp. *hermaphroditum* make up the bulk of the mat with scattered *Vaccinium myrtillus* and occasional *V. vitis-idaea*, *Arctostaphylos uva-ursi*, *Juniperus communis nana* and *Salix herbacea*. *Deschampsia flexuosa* occurs sparsely with the above herbs and other more occasional species. Small patches of *Racomitrium lanuginosum* are frequent and there is a rich and extensive lichen flora including the species listed above.

H17b

***Empetrum nigrum* ssp. *nigrum* sub-community**

Calluna vulgaris dominates the sub-shrub mat with *Arctostaphylos alpinus* constant and *Empetrum nigrum* ssp. *nigrum* and *Erica cinerea* as common associates. *Vaccinium myrtillus* is again sparse with occasional *Arctostaphylos uva-ursi*. *Carex bigelowii* is less common and *Deschampsia flexuosa* more frequent than in H17a, with the other associates listed above. Among bryophytes *Hypnum jutlandicum* frequently joins *Racomitrium lanuginosum*. Lichens are not so varied or abundant as in H17a, but *Cladonia uncialis* and *C. arbuscula* remain very common.

H18 *Vaccinium myrtillus –*
Deschampsia flexuosa heath

This community includes a variety of moss-rich and grassy sub-shrub vegetation, in which *Vaccinium myrtillus* is the most frequent ericoid, with *Calluna vulgaris* only occasional and often lacking in vigour. Other sub-shrubs can make a sizeable contribution to the canopy; in particular *Empetrum nigrum*, usually ssp. *hermaphroditum*, is most frequent, often forming patches. *Vaccinium vitis-idaea* is also common.

Among vascular associates *Deschampsia flexuosa* and *Galium saxatile* are constant throughout, with *Nardus stricta*, *Agrostis canina* ssp. *montana* and *Potentilla erecta* all very frequent. In some stands these species account for virtually all the herbaceous cover. The grasses *Festuca ovina*, *Agrostis capillaris* and *Anthoxanthum odoratum* occur at least occasionally and increase in frequency and abundance in some sub-communities.

The other element, which is usually prominent, comprises bulky mosses. *Dicranum scoparium*, *Pleurozium schreberi* and *Hypnum cupressiforme s.l.* are very common throughout, *Hylocomium splendens* is also conspicuous and there can be frequent *Rhytidiadelphus loreus*, *R. squarrosus*, *Plagiothecium undulatum*, *Dicranum majus* and *Racomitrium lanuginosum*. Some lichens occur frequently although an extensive carpet is never found. The most common species are *Cladonia arbuscula*, *C. impexa* and *C. uncialis*.

This community is typical of moist but free-draining, base-poor to circumneutral soils over steeper slopes at moderate to high altitudes through the uplands of northern Britain. It is largely confined to altitudes above 400 m and can extend up to 800 m. It occurs over a wide variety of bedrocks on a variety of soil profiles which have a superficial pH of 3.5-5.5. Typically, however, the soils have a well-developed humic layer. This community is widespread through the uplands of Britain but is particular common in northern Scotland, where the heart of its range occurs in the central and eastern Highlands, with more sporadic occurrences to the north-west.

At higher levels this vegetation is probably a natural climax with the floristics and distribution of the community being influenced by snow-lie, but towards the sub-montane zone quite extensive stands of the vegetation have been biotically derived as a result of woodland clearance and pasturing. In other places treatments such as burning and grazing have precipitated its spread on to blanket peats.

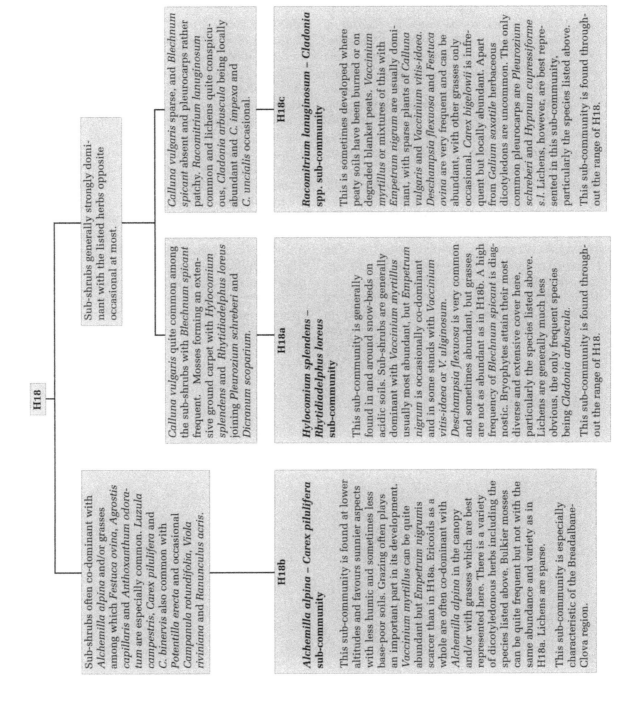

H18

Sub-shrubs often co-dominant with *Alchemilla alpina* and/or grasses among which *Festuca ovina*, *Agrostis capillaris* and *Anthoxanthum odoratum* are especially common. *Luzula campestris*, *Carex pilulifera* and *C. binervis* also common with *Potentilla erecta* and occasional *Campanula rotundifolia*, *Viola riviniana* and *Ranunculus acris*.

Sub-shrubs generally strongly dominant with the listed herbs opposite occasional at most.

Calluna vulgaris quite common among the sub-shrubs with *Blechnum spicant* frequent. Mosses forming an extensive ground carpet with *Hylocomium splendens* and *Rhytidiadelphus loreus* joining *Pleurozium schreberi* and *Dicranum scoparium*.

Calluna vulgaris sparse, and *Blechnum spicant* absent and pleurocarps rather patchy. *Racomitrium lanuginosum* common and lichens quite conspicuous, *Cladonia arbuscula* being locally abundant and *C. impexa* and *C. uncialis* occasional.

H18b

Alchemilla alpina – Carex pilulifera sub-community

This sub-community is found at lower altitudes and favours sunnier aspects with less humic and sometimes less base-poor soils. Grazing often plays an important part in its development. *Vaccinium myrtillus* can be quite abundant but *Empetrum nigrumis* scarcer than in H18a. Ericoids as a whole are often co-dominant with *Alchemilla alpina* in the canopy and/or with grasses which are best represented here. There is a variety of dicotyledonous herbs including the species listed above. Bulkier mosses can be quite frequent but not with the same abundance and variety as in H18a. Lichens are sparse.

This sub-community is especially characteristic of the Breadalbane-Clova region.

H18a

Hylocomium splendens – Rhytidiadelphus loreus sub-community

This sub-community is generally found in and around snow-beds on acidic soils. Sub-shrubs are generally dominant with *Vaccinium myrtillus* usually most abundant, but *Empetrum nigrum* is occasionally co-dominant and in some stands with *Vaccinium vitis-idaea* or *V. uliginosum*. *Deschampsia flexuosa* is very common and sometimes abundant, but grasses are not as abundant as in H18b. A high frequency of *Blechnum spicant* is diagnostic. Bryophytes attain their most diverse and extensive cover here, particularly the species listed above. Lichens are generally much less obvious, the only frequent species being *Cladonia arbuscula*.

This sub-community is found throughout the range of H18.

H18c

Racomitrium lanuginosum – Cladonia spp. sub-community

This is sometimes developed where peaty soils have been burned or on degraded blanket peats. *Vaccinium myrtillus* or mixtures of this with *Empetrum nigrum* are usually dominant, with sparse plants of *Calluna vulgaris* and *Vaccinium vitis-idaea*. *Deschampsia flexuosa* and *Festuca ovina* are very frequent and can be abundant, with other grasses only occasional. *Carex bigelowii* is infrequent but locally abundant. Apart from *Galium saxatile* herbaceous dicotyledons are uncommon. The only common pleurocarps are *Pleurozium schreberi* and *Hypnum cupressiforme* s.l. Lichens, however, are best represented in this sub-community, particularly the species listed above.

This sub-community is found throughout the range of H18.

H19 *Vaccinium myrtillus –* *Cladonia arbuscula* heath

This community consists essentially of a very low mat, 5-10 cm high, of sub-shrubs with an abundance of lichens, often marking stands with a yellowish tinge. Lichens are more extensive and dominant than in the *Calluna vulgaris – Cladonia arbuscula* heath (H13). *Calluna vulgaris* is uncommon overall and *Vaccinium myrtillus* is the most consistent sub-shrub, being co-dominant in more sheltered situations, although sparse in exposed sites. *Vaccinium vitis-idaea* is less common, but constant, and *V. uliginosum* scarce overall. *Empetrum nigrum*, almost always ssp. *hermaphroditum*, is frequent and can exceed *Vaccinium* species in its cover.

Vascular associates are few but *Carex bigelowii*, a constant, is frequent and often abundant and may be co-dominant with the ericoids and lichens. The other common and constant plant is *Deschampsia flexuosa*. *Festuca ovina* is also fairly frequent together with several herbs, such as *Galium saxatile*, in sub-community H19a.

Bryophytes are generally not important and *Racomitrium lanuginosum* is only abundant in one sub-community. *Dicranum fuscescens* is quite frequent or there may only be sparse shoots of *Polytrichum alpestre* and *P. piliferum*.

Much more important are the lichens, particularly larger fruticose species such as *Cladonia arbuscula* and *C. uncialis*, both constants, and less commonly *C. rangiferina* and *C. gracilis*, mixtures of which can exceed the sub-shrubs in total cover. *Cetraria islandica* and *Cornicularia aculeata* are also very common, often with a variety of other species.

This heath is typical of base-poor soils on moderately sheltered and snow-bound slopes at high altitudes, particularly in the more continental mountains of northern Britain. The vegetation is strongly montane, being found mainly above 650 m up to 1000 m or even beyond, and usually stands are found in sites with some shelter so that there is winter protection from lying snow. It is characteristic of acid soils with a superficial pH of about 4. It has a similar range to that of H13, being strongly concentrated in the central and eastern Highlands of Scotland, particularly the Grampians, but also in the mountains of the north-west and the Southern Uplands and with scattered localities in northern England.

Floristic variation reflects differences in exposure and soil type, but overall the vegetation is a climatic climax.

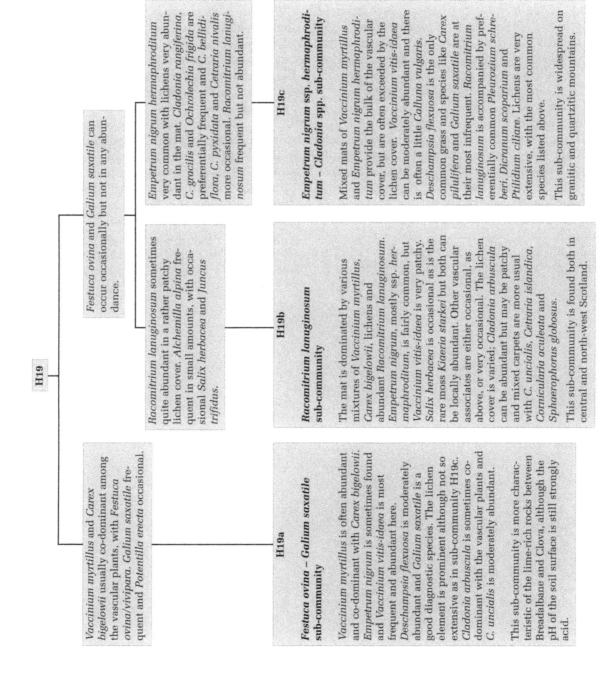

H19

Vaccinium myrtillus and *Carex bigelowii* usually co-dominant among the vascular plants, with *Festuca ovina/vivipara. Galium saxatile* frequent and *Potentilla erecta* occasional.

Festuca ovina and *Galium saxatile* can occur occasionally but not in any abundance.

Racomitrium lanuginosum sometimes quite abundant in a rather patchy lichen cover. *Alchemilla alpina* frequent in small amounts, with occasional *Salix herbacea* and *Juncus trifidus.*

Empetrum nigrum hermaphroditum very common with lichens very abundant in the mat. *Cladonia rangiferina, C. gracilis* and *Ochrolechia frigida* are preferentially frequent and *C. bellidiflora, C. pyxidata* and *Cetraria nivalis* more occasional. *Racomitrium lanuginosum* frequent but not abundant.

H19a

Festuca ovina – Galium saxatile sub-community

Vaccinium myrtillus is often abundant and co-dominant with *Carex bigelowii. Empetrum nigrum* is sometimes found and *Vaccinium vitis-idaea* is most frequent and abundant here. *Deschampsia flexuosa* is moderately abundant and *Galium saxatile* is a good diagnostic species. The lichen element is prominent although not so extensive as in sub-community H19c. *Cladonia arbuscula* is sometimes co-dominant with the vascular plants and *C. uncialis* is moderately abundant.

This sub-community is more characteristic of the lime-rich rocks between Breadalbane and Clova, although the pH of the soil surface is still strongly acid.

H19b

Racomitrium lanuginosum sub-community

The mat is dominated by various mixtures of *Vaccinium myrtillus, Carex bigelowii,* lichens and abundant *Racomitrium lanuginosum. Empetrum nigrum,* mostly ssp. *hermaphroditum,* is fairly common, but *Vaccinium vitis-idaea* is very patchy. *Salix herbacea* is occasional as is the rare moss *Kiaeria starkei* but both can be locally abundant. Other vascular associates are either occasional, as above, or very occasional. The lichen cover is varied; *Cladonia arbuscula* can be abundant but may be patchy and mixed carpets are more usual with *C. uncialis, Cetraria islandica, Cornicularia aculeata* and *Sphaerophorus globosus.*

This sub-community is found both in central and north-west Scotland.

H19c

Empetrum nigrum ssp. hermaphroditum – Cladonia spp. sub-community

Mixed mats of *Vaccinium myrtillus* and *Empetrum nigrum hermaphroditum* provide the bulk of the vascular cover, but are often exceeded by the lichen cover. *Vaccinium vitis-idaea* can be moderately abundant and there is often a little *Calluna vulgaris. Deschampsia flexuosa* is the only common grass and species like *Carex pilulifera* and *Galium saxatile* are at their most infrequent. *Racomitrium lanuginosum* is accompanied by preferentially common *Pleurozium schreberi, Dicranum scoparium* and *Ptilidium ciliare.* Lichens are very extensive, with the most common species listed above.

This sub-community is widespread on granitic and quartzitic mountains.

H20 *Vaccinium myrtillus – Racomitrium lanuginosum* heath

This community brings together a variety of vegetation types in which *Vaccinium myrtillus* and/or *Empetrum nigrum* ssp. *hermaphroditum* occur, occasionally with other sub-shrubs such as *V. vitis-idaea*, and are co-dominant with *Racomitrium lanuginosum* or hypnaceous mosses. *Vaccinium myrtillus* and *E. nigrum* form a low mat, usually less than 10 cm high, appearing as a patchy mosaic of bushes among the moss carpet. At lower altitudes *Juniperus communis* ssp. *nana* and *Erica cinerea* can show local prominence.

Among vascular associates *Carex bigelowii*, *Festuca ovina/vivipara*, *Deschampsia flexuosa* and *Galium saxatile* are all constant and frequent. There are few other common herbs, although the grasses may include frequent *Nardus stricta*. *Huperzia selago* and *Potentilla erecta* are frequent in some stands and may be accompanied by *Thymus praecox*, *Viola riviniana* and *Carex pilulifera*.

Much of the distinctive character of this vegetation type depends on the cryptogams. *Racomitrium lanuginosum* is very important, forming a woolly carpet, and it is found with a variety of other bulky mosses. *Hypnum cupressiforme s.l.*, *Hylocomium splendens*, *Rhytidiadelphus loreus* and *Pleurozium schreberi* are all constant and can be prominent. Additionally, *Polytrichum alpinum* and *Dicranum scoparium* are found in many stands. Common hepatics are *Ptilidium ciliare* and *Diplophyllum albicans*, but their greatest variety is found in the *Bazzania – Mylia* sub-community. Lichens are less important, but *Cladonia uncialis* and *C. arbuscula* are most frequent throughout and may be modestly abundant, with *C. gracilis* and *Cetraria islandica* also common.

This heath is characteristic of humic, base-poor soils on fairly exposed slopes and summits at moderate to high altitudes, in the cool oceanic mountains of north-west Scotland, extending to Skye, and scattered through the Grampians. Almost always, the bedrocks underlying this heath are siliceous in character.

Climatic differences and some modest variation in edaphic conditions influence the floristics, but this is essentially climax vegetation.

H20

Vaccinium vitis-idaea very scarce, but Calluna vulgaris, Erica cinerea and Juniperus communis nana occasional with Alchemilla alpina sometimes co-dominant. Potentilla erecta, Thymus praecox, Viola riviniana and Carex pilulifera frequent.

Vaccinium vitis-idaea very common and all other listed associates scarce.

Rich and luxuriant patchwork of bryophytes present among the Racomitrium lanuginosum carpet with Dicranum scoparium, Plagiothecium undulatum, Sphagnum capillifolium, Mylia taylori, Diplophyllum albicans, Pleurozia purpurea, Bazzania tricrenata, Scapania gracilis. S. ornithopodioides, Anastrepta orcadensis and Anthelia julacea frequent.

Diplophyllum albicans and Anastrepta orcadensis sometimes found but combinations of these other bryophytes rare.

Racomitrium lanuginosum reduced in cover with dominance often passing to mixtures of Pleurozium schreberi, Hylocomium splendens and Rhytidiadelphus loreus.

Bulky pleurocarps can be common but not abundant among the dominant Racomitrium lanuginosum carpet, but the lichen carpet is richer with frequent Cladonia gracilis, Cetraria islandica and Cornicularia aculeata, and occasional Cladonia leucophaea. Sphaerophorus globosus and Alectoria nigricans.

H20a

Viola riviniana – Thymus praecox sub-community

The sub-shrub mat is more varied than usual in the community. Empetrum nigrum hermaphroditum and Vaccinium myrtillus can both be abundant, but Alchemilla alpina can be co-dominant and more locally Juncus communis ssp. nana or Erica cinerea may occur with sparse shoots of Calluna vulgaris. Herbs characteristic of H20, such as Festuca ovina/vivipara, Deschampsia flexuosa and Galium saxatile, remain very frequent and are joined preferentially by the herbs listed above. The cryptogam flora is poor, with Racomitrium lanuginosum as the typical dominant, but other pleurocarps are less frequent and Diplophyllum albicans is the only common hepatic.

This sub-community extends the range of H20 into the milder foothills of the western seaboard and Skye.

H20c

Bazzania tricrenata – Mylia taylori sub-community

Here the general floristic features are as in H20b though Blechnum spicant and Juncus trifidus are frequent. The community mosses are well represented but distinctively there is a range of conspicuous bryophytes, including Atlantic hepatics as listed above. Lichens are generally sparse.

This sub-community is very much confined to the wettest regions and even then it is restricted to suitable cold and damp aspects.

H20d

Rhytidiadelphus loreus – Hylocomium splendens sub-community

Although Racomitrium lanuginosum is generally reduced in cover, the general vegetation features accord well with the community as a whole. Vaccinium vitis-idaea is preferentially common, Carex bigelowii and Deschampsia flexuosa are frequent and Galium saxatile and Festuca ovina/vivipara are more occasional. The moss mat is distinctive with dominance passing to the species listed above. The Atlantic hepatics found in H20c are absent.

This sub-community is local throughout the range of H20.

H20b

Cetraria islandica sub-community

Mixtures of Empetrum nigrum ssp. hermaphroditum and Vaccinium myrtillus with abundant Racomitrium lanuginosum are usually dominant. Grasses like Festuca ovina/vivipara. Deschampsia flexuosa. Nardus stricta, Agrostis canina and Anthoxanthum odoratum tend to be prominent, but the richness of vascular plants in H20a is absent. Bryophytes are not numerous although community constants are all common with frequent Hypnum cupressiforme s.l. and Polytrichum alpinum. The lichen flora is a little richer than usual with the species listed above present.

This is the most widespread and common sub-community overall.

H21 *Calluna vulgaris – Vaccinium myrtillus – Sphagnum capillifolium* heath

This community has a mixed canopy of sub-shrubs, usually 30-50 cm high, with a damp layer of luxuriant bryophytes. *Calluna vulgaris* is usually the dominant ericoid, although *Vaccinium myrtillus* is constant and *Empetrum nigrum*, almost always ssp. *hermaphroditum*, very frequent. *Erica cinerea* is also frequent, but patchy. Other sub-shrubs are only occasional.

Deschampsia flexuosa* and *Potentilla erecta* are constant and very common though usually present as sparse, scattered individuals. More distinctively *Blechnum spicant* is constant and *Solidago virgaurea* and *Listera cordata* frequent. There are only occasional records for other vascular associates.

The bryophytes form an extensive and lush carpet. Constant throughout are bulky hypnaceous mosses such as *Hypnum cupressiforme s.l.*, *Rhytidiadelphus loreus*, *Pleurozium schreberi* and *Hylocomium splendens*, with *Plagiothecium undulatum*, *Dicranum scoparium* and *D. majus* also very common. Particularly distinctive is the high frequency and local abundance of *Sphagnum capillifolium*. *Racomitrium lanuginosum* becomes more frequent at higher altitudes. The most spectacular enrichment in this element comes from oceanic hepatics and this community is a major locus for the 'mixed northern hepatic mat'. Species such as *Scapania gracilis*, *Mylia taylori* and *Diplophyllum albicans* can be found throughout, but the *Mastigophora – Herbertus* sub-community has an additional range of Atlantic species, forming a unique vegetation found at higher altitudes in north-west Scotland where summer temperatures are lower and rainfall higher. Lichens are fairly insignificant, *Cladonia impexa* being the only species occurring commonly throughout.

This heath is highly characteristic of fragmentary humic soils, developed in situations with a cool but equable climate and a consistently shady and extremely humid atmosphere. It is almost wholly confined to low to moderate altitudes through the oceanic mountains of north-west Scotland and on Skye, with outliers on Orkney, in south-west Scotland and the Lake District.

It is largely restricted to steep, sunless slopes of north-west to easterly aspect, often with rock outcrops and blocky talus, among which crevices provide additional shade. In some situations this may not be a climax community but a result of woodland clearance, but towards the upper end of its altitudinal limits this heath appears to form a natural component of vegetation patterns controlled largely by variations in local climates and soils. It is sometimes lightly grazed, but burning is very damaging and recovery is probably extremely slow. It seems certain that the extent of this community has been reduced by burning.

H21

Empetrum nigrum hermaphroditum frequent and locally abundant among the sub-shrubs with especially rich and luxuriant cryptogam carpets among which there is frequent *Racomitrium lanuginosum, Mylia taylori, Scapania gracilis, Bazzania tricrenata, Pleurozia purpurea, Diplophyllum albicans, Anastrepta orcadensis, Mastigophora woodsii, Herbertus aduncus hutchinsiae, Cladonia uncialis* and *C. arbuscula.*

Empetrum nigrum hermaphroditum local and combinations of listed cryptogams rare, but *Dicranum scoparium* common with frequent fronds of *Pteridium aquilinum.*

H21b

Mastigophora woodsii – Herbertus aduncus ssp. hutchinsiae sub-community

Calluna vulgaris is usually the most abundant sub-shrub, but the canopy is short and more mixed than in H21a. The bryophytes are extremely well developed. Among the mosses all the community constants occur frequently. The hepatics, however, are most abundant, tingeing the vegetation with a variety of colours. They include the species listed above with other rarer Atlantic hepatics.

This sub-community is restricted in range, being confined to the more shaded and humid habitats in north-west Scotland.

H21a

Calluna vulgaris – Pteridium aquilinum sub-community

This sub-community occurs in sites which cannot support the full range of hepatics. *Calluna vulgaris* is generally a strong dominant in this taller and more species-poor heath. *Vaccinium myrtillus* is very common with *Erica cinerea* and *Vaccinium vitis-idaea* occasional. Other vascular plants are sparse, but distinctive is *Pteridium aquilinum* with occasional *Oxalis acetosella, Viola riviniana* and *Luzula sylvatica.* Bryophytes can have fairly high cover, but comprise almost entirely the community constants.

This sub-community is found throughout the range of H21.

H22 *Vaccinium myrtillus – Rubus chamaemorus* heath

This heath has a mixed cover of sub-shrubs over a moist cover of bryophytes similar to that of *Calluna vulgaris – Vaccinium myrtillus – Sphagnum capillifolium* heath (H21). However, here the canopy is not as tall, being mostly between 10 and 30 cm high, and *Calluna vulgaris* is not invariable in its dominance (*Vaccinium myrtillus* is dominant in the *Polytrichum – Galium* sub-community). *Empetrum nigrum* ssp. *hermaphroditum* is constant, as is *V. vitis-idaea* (although less frequent), and *V. uliginosum* is rare. *Erica cinerea* is absent.

The vascular associates are distinctive because, with constant *Deschampsia flexuosa*, there is frequently a little *Rubus chamaemorus* and *Cornus suecica*. *Eriophorum vaginatum* can be locally abundant and there are records for *Potentilla erecta*, *Melampyrum pratense*, *Listera cordata*, *Juncus squarrosus* and *Nardus stricta*.

Bryophytes are always conspicuous and sometimes very abundant. *Dicranum scoparium* and the hypnaceous mosses *Pleurozium schreberi*, *Hylocomium splendens* and *Rhytidiadelphus loreus* are the most consistent and constant, although *Sphagnum* spp. can also have a high cover, with the constant *S. capillifolium* being especially common and several other species locally abundant. A variety of other mosses and hepatics are variable in their occurrence. Lichens are typically less prominent, although *Cladonia arbuscula* is constant and can show modest abundance.

This heath is characteristic of wet, base-poor peats at moderate to high altitudes (mainly between 500 m and 800 m), where there is protection against extremes of dryness and winter cold by virtue of an oceanic influence or locally prolonged snow-lie. The profiles found beneath this community are typically poorly-developed, often consisting of just a layer of bryophyte or ericoid humus resting directly on blocky talus, derived from a variety of pervious bedrocks. It is almost entirely confined to the central and north-west Highlands of Scotland. In the former region it is typical of early snow-beds where it is mainly present as the *Polytrichum – Galium* sub-community. In the north-west Highlands, where the climate is ameliorated by the oceanic climate, this heath is generally represented by the *Plagiothecium – Anastrepta* sub-community.

Climatic and edaphic factors maintain this heath as a climax vegetation in most situations, although at its lowest limits it falls within the altitudinal range of historical pine forest. It is sometimes affected by grazing and burning where these treatments are applied to the surrounding heaths. Burning is deleterious to the floristic richness of the community.

H22

Carex bigelowii frequent in small amounts with a rich and extensive patchwork of cryptograms, among which *Racomitrium lanuginosum, Plagiothecium undulatum, Ptilidium ciliare, Anastrepta orcadensis, Barbilophozia floerkii, Cladonia bellidiflora, C. uncialis, C. leucophaea, C. gracilis* and *C. impexa* are very common.

Carex bigelowii and cryptograms listed opposite all very scarce, but *Gallium saxatile* and *Blechnum spicant* frequent as scattered individuals and *Polytrichum commune* very common among usually plentiful hypnaceous mosses.

H22b

Plagiothecium undulatum – Anastrepta orcadensis sub-community

Calluna vulgaris is often a strong dominant in a taller canopy with *Empetrum nigrum* ssp. *hermaphroditum* occasionally abundant. *Vaccinium myrtillus* usually has low cover and *V. vitis-idaea* occurs unevenly. *Cornus suecica, Rubus chamaemorus* and *Deschampsia flexuosa* are all more patchy than in H22a, though *Eriophorum vaginatum* is common as scattered shoots. *Carex bigelowii* and *Huperzia selago* are preferential at low frequencies. The cryptogams are distinctive with both hypnaceous mosses and *Sphagnum* spp. prominent together with a number of Atlantic hepatics including the species listed above. Lichens are also more numerous in this sub-community with the above species present and *Cladonia arbuscula* constant.

This sub-community occurs throughout the range of H22 and is particularly well-developed in the north-west Highlands.

H22a

Polytrichum commune – Galium saxatile sub-community

Vaccinium myrtillus is generally dominant in a low sub-shrub canopy with *Calluna vulgaris* and/or *Empetrum nigrum* ssp. *hermaphroditum* sub-dominant. *Vaccinium vitis-idaea* is fairly common but of low cover. *Cornus suecica* and *Rubus chamaemorus* are most consistently frequent here though not abundant. *Deschampsia flexuosa* is rather patchy and *Blechnum spicant* and *Galium saxatile* are preferentially common. Hypnaceous mosses, especially *Hylocomium splendens* and *Rhytidiadelphus loreus*, are plentiful with *Sphagnum capillifolium* patchily abundant. Lichens are rare apart from scattered *Cladonia arbuscula*.

This sub-community is largely restricted to the central Highlands.

CPSIA information can be obtained
at www.ICGtesting.com
Printed in the USA
BVHW01s0459270318
511680BV00002B/11/P